MAR 2 9 2004

Hand picked by the author, this book contains the best articles from his years of newspaper writing.

Included in this volume are pollen and mold data from cities around the country, scores of articles about indoor air quality, pets, chemicals in the home, allergies, asthma, buying and selling a home, plus a special essay by the author, "Does Television Cause Asthma?"

A Breath of Fresh Air

Mark R. Sneller, Ph.D.

Fresh Air Press Tucson, Arizona

Copyright © 2003 by Mark R. Sneller, Ph.D.

All rights reserved. No part of this book may be reproduced or transmitted in any form or by any means, electronic or mechanical, including photocopying, recording, or by any information storage and retrieval system, without permission in writing from the author.

This edition was prepared for printing by
Ghost River Images
5350 East Fourth Street
Tucson, Arizona 85711
www.ghostriverimages.com

Cover illustration by: Tina Spengler

Edited by: Jim Woods

ISBN 0-9741131-0-7

Library of Congress Control Number: 2003106043

Printed in the United States of America

Second Printing: July, 2003

10 9 8 7 6 5 4 3 2

This book is dedicated to my mother
who gave me my first books to read and
taught me to persevere.

When I once told her that I would be an
old man by the time I got out of school,
she replied, "You're going to be an old
man anyway, so you might as well be a
smart old man."

Contents

INTRODUCTION

Why this book is important to everyone.

This book covers virtually every aspect of the countless particles that we breathe and where we come in contact with them. It discusses the irritating gases that we are told to believe will help us, and in fact, have just the opposite effect.

Yes, our vitality and that of our children is at stake. Think of yourself trying to function, trying to maintain a positive attitude on a smoggy day, or in a freshly painted room or a home strongly scented with formaldehyde.

Let A Breath of Fresh Air be your guidebook, your encyclopedia.

A Breath of Fresh Air is a collection newspaper columns written over the past several years and read by a half-million persons weekly. The columns are written for the layperson and are based upon my twenty-five years experience as the owner of an indoor air quality company (Aero-Allergen Research, LLC). They also are based on facts gleaned from the reading of hundreds of publications in scientific journals as well as books. The more salient publications are listed in the selected reading portion of this book. Many of these are several years old and provide the basis for much of today's knowledge in their representative fields.

Best of all, this book applies to everyone. Where there exists controversy, that controvery will be discussed. And wherever you live, you will find valuable information in these volumes.

We have all been witness to a great number of technological changes in the last few decades. In many ways this has not been for the good. Nowhere is this more true than in the field of respiratory health, which pertains to every person who breathes, not just those with allergy and asthma. This book tells us about what happened and best of all, it tells us what we can do about it—at little or no extra cost.

Doctors and scientists tend to place the blame on the shoulders of tight buildings, which resulted from the energy crisis of the seventies. Yes, ninety percent of "sick-building" problems are due to a lack of fresh air. But it's more than that. It has to do with the fact that there are no regulations regarding the sales of products to the home dweller. It's a seller's market. It has to do with our gullibility with only a few strong voices in the wilderness speaking for simplicity.

In short, our ability to breathe safely affects our thought process and our behavior toward ourselves and others. Nothing complicated here. The blood goes to the brain.

This book presents the basics of allergy and asthma, the chemicals in our everyday lives and their safe and inexpensive substitutes. It presents basic information on the home or apartment, from buying or selling or renting, to keeping it free from respiratory irritants at little or no extra cost. Do you want to know the latest information about the ozone machines or any type of "air cleaning" device? Read on.

I'll leave you with this. Small challenges can add up to one large challenge. Allergens added to fragrances added to chemical smells added to diesel exhaust added to pesticides will cause the cup to run over. Each of these saps our energy, our vitality and our sexual desire. Together they work in synergism, where the world of one plus one equals ten. Elimination and simplification doesn't require money. It only requires education.

Mark R. Sneller, Ph.D.

CHAPTER 1
ALLERGY BASICS

ALLERGY Q AND A

Q. What is an allergen?

A. An allergen is a substance that causes the production of a certain class of antibodies that can lead to an allergic reaction. There are a lot of things we know today about what it takes to make an allergen that we didn't know a few years ago. For example, every single protein has the ability to be come an allergen.

Q. What is an antigen?

A. Antigens are chemical structures. They turn a chemical substance such as a protein into an allergen. Thanks to the antigen, this allergen then can elicit an allergic reaction. One way to remember it is that an allergen generates allergies and antigens generate antibodies.

Q. What does sensitization to allergens mean?

A. To become sensitized means that the immune response becomes primed to an allergen(s); it becomes exposed for the first time. It is the first step in an allergic attack and often occurs during infancy, when the immune system is not fully developed. One can still become sensitized at any time in life. Also, by itself, formaldehyde and other chemicals may not be harmful other than being irritants. However, they can act as a priming agent and sensitize the body to react to allergens.

Q. Suppose we are exposed to several allergens such as pollen, mold, cat allergen and latex. Are we more sensitive to allergens in general now that we have been exposed to more of them?

A. Yes. The more negative factors that are involved then the more sensitive we become. Negative factors include: more frequent and longer exposure to antigens, fatigue, heredity and weakened immune response.

The immune response can be weakened by a number of factors such as illness, depression, stress, poor nutrition and exposure to pesticides, steroids and chemotherapy.

Also, the type of allergen is important. For example, many more powerful allergens from pollen grains are released very quickly in the body; other weaker allergens are released by other pollen types more slowly. Those that are released more quickly can cause a rapid response by the immune system.

Q. What is a cross-reaction?

A. A cross-reaction occurs when antigens from different substances have a chemical structure that is similar. They share common antigens. For example, if you are allergic to birch pollen you can also react to latex, celery, banana, avocado and mugwort. If you are sensitized to proteins that are found in some pollen types, this could be a major reason for development of food allergies associated with vegetables.

Q. How does air pollution affect the allergic response?

A. Ozone and sulfur dioxide are two pollutants associated with automobile exhaust. Latex is a pollutant associated with rubber tires. The first two enhance your sensitivity to allergens in general. Antigens found in latex are found in many foods and pollens. Also new studies are finding that pollutant gases tend to coat pollen grains to make them more potent.

ALLERGY MYSTERIES

Did you ever wonder how you could have a respiratory response to allergens and you could swear you were taking every precaution? Allergy need not be so mysterious when we understand it better.

Cat antigen: More homes have cat antigen in their air than have cats. It attaches to shoes and clothing from an area where there was or is a cat and taken to an area where there is none. Also, even though you may have given up your cat, it takes months and even years for the antigen to disappear from the soft furnishings and mattresses. Bird antigen is also long lived.

To my knowledge, there have been no studies that link library books with transmission of cat antigen or any other antigen. However, it is reasonable to assume that a book will pick up antigen and transmit it. This is specially true if it is laying around in an environment where say, cat, dust mite and cockroach antigens are present. Books are intimate objects that we handle and take to bed with us. Maybe it's not just the dust on books that makes us sneeze.

Latex: You got rid of your powdered latex gloves but still have itching and watering eyes as if you hadn't. And now you have rashes as well. If you live near a busy intersection you could be breathing in the latex particles that are shed from rubber tires. Also, people who are sensitive to latex are frequently reactive to banana, avocado, kiwi, chestnut and birch; hence the rashes.

Sneeze upon waking: It's not pollen because the doctor told you to sleep with the windows closed. In fact, you may have come in contact with an allergen before going to bed. It is not until four o'clock in the morning that natural biorhythms and resistance become lowered enough for bodily reactions to occur.

Do you sneeze upon going outdoors? Bright sunlight triggers photo-receptors in the eye which activates nerves in the nose. It may not be allergies at all.

Dust mites: Your skin tests show you are allergic to house dust mites. So you lower the indoor humidity to less than forty-five percent to get rid of house dust mites but you are reactive just the same. In this case the dust mites became dormant to wait it out. But their feces is still present. That's what triggers the reaction.

Food allergy: You know you are allergic to wheat. That's why you don't eat it but let the rest of the family have all they want. You are still sneezing and have a sinus headache in the house. It's time to do some serious dusting and make the kids eat at the table. Studies show that food antigens are present in the dust we breathe. In some cases food antigens are in the smell of cooked food and in food particles that have fallen to the floor, to be ground underfoot and mixed with dust.

To complicate matters, many antigens that are present in some foods are also present in other foods and other substances. For example, antigens found in birch pollen are also found in tomatoes, bananas and kiwi fruit.

HAY FEVER

Allergic rhinitis, hay fever or common allergy is so common that for men in the U.S. work force it is more common than back problems. For women it is more common than back problems or hypertension.

Allergic rhinitis is different from asthma. It is not life threatening and dollar-wise it is a low cost disease. It's symptoms are nasal itching and congestion, sneezing and red and itching eyes with tearing. Quite often the patient has the feeling of general fatigue and just doesn't feel well overall.

The employee with this illness takes a sick day off or does not produce work at his or her best level. Loss of sleep from the illness also makes for poor work performance. It is ironic, but it is the inexpensive medical care that is part of the problem. For fifty years people have been able to get over-the-counter anti-histamines. These produce drowsiness and make us mentally and physically slow. About half the working men and women who have hay fever use these over-the-counter anti-histamines. The cure may be worse than the disease.

In one study it was found that some ten percent of working men and twelve percent of working women had allergic rhinitis. This translates to thirteen million workers. With millions of sleepy men and women trying to work we produce less product at a lower quality as a nation.

An almost identical problem exists with children who have the disease. If the disease is severe then they may be absent from school, or may be irritable or tired from loss of sleep. As a result of this learning suffers. With the right medication the symptoms can be cured. With the wrong medication the ability to learn in and out of school could become worse. One would think we would know something as simple as this. But as of this date none of the hay fever medications currently available have been studied in terms of their effect on learning in the young student. This includes such medications as glucocorticoids, decongestants, and anti-histamines. The reason for this is that their primary purpose is to cure a clinical problem and their ability to cure that problem is what is studied most.

New medications such as non-drowsy or non-sedating anti-histamines may not have such a negative effect. They are being used more often with working men and women and with children as well. This is the new generation of medicines for allergic rhinitis. We can see why it is very important for the parent, teacher and school nurse to be aware of the fact that the child's slow learning may be due to hay fever. The child's behavior may be affected by one particular medication, but not another.

COMMON ALLERGENS

Feathers: The most common source of feathers in the home is the pillow (down). The best pillow for an allergic individual is made of Acrilan or Dacron. These will last a lifetime with proper care. They should be covered with a dust proof cover to prevent dust rom settling into the stuffing. These covers can be obtained from drug stores or department stores. Obviously, the presence of birds in the home of allergic patients should be discouraged.

Wool: Woolens are to be avoided and may contain pesticide.

Cottonseed: Most inhalant allergy traced to cotton is due to particles of cottonseed. These are commonly found in the stuffing of pillows, sofa cushions, mattresses, bed pads, blankets, and furniture. If these items cannot be eliminated they should be covered with dust-proof covers. Most miniature golf courses are covered with ground cottonseed mixed with other substances.

Cottonseed is also found in some cattle and poultry feeds. While cotton may still have residue of pesticide, cotton fabrics are commonly treated with formaldehyde and other respiratory irritants for permanent- press and fireproofing purposes.

Kapok: This is a plant product related to cotton. It is most often found in pillows, sofa cushions, mattresses and comforters. All kapok crumbles to dust in the course of a couple of years and should be avoided.

Flaxseed: Flaxseed can cause allergic reactions after being eaten, inhaled or brought into contact with the skin. Sources include certain cereals such as milk from cows fed flaxseed, muffins, laxatives, certain cough medicines, wave sets, shampoos, oil cloth, wax, patent leather, paints and varnishes, putty, soft soap, linen and other products.

Pyrethrum: This is the dried flower of the pyrethrum plant, a member of the chrysanthemum family. Chrysanthemums and Pyrethrum are closely related to ragweed. Thus, ragweed sensitive patients may also experience symptoms upon exposure to pyrethrum. This is a common ingredient in insect powders and sprays for the house, garden and pets. Pyrethrum is in many insecticides. It is very popular for use in mothproofing carpets, draperies and upholstery and to prevent the growth of insects in these materials. Thus, it is advisable that the pyrethrum-sensitive patient not go into rooms or closets which have been recently treated with the insecticide, or wear clothes which have been recently mothproofed. This substance is also found in ointments. Pyrethrum sensitivity may be seasonal and hyposensitization by the allergist may be required.

Orris root: Orris is obtained by powdering the root of a plant related to the iris. It is often used in cosmetics, scented soaps, toothpowders, bath salts, perfumes, shaving creams and lotions.

Sulfa compounds: Many insecticides that are used out of doors contain sulfa compounds. Persons who are allergic to sulfa may react to these compounds when they are inhaled.

CHAPTER 2
ASTHMA

WHAT IS ASTHMA?

According to the United States Department of Health and Human Services, along with its sister allergic diseases, bronchial asthma is among the most common chronic diseases suffered by Americans. Approximately 15 to 16 million Americans suffer from bronchial asthma; between 30 and 35 million have other allergic diseases. In other words, these are very common diseases.

Asthma is often thought to be a lingering disease, but about 5,000 Americans die from asthma each year. The group at greatest risk is older asthmatics, above the age of 50.

Asthma is not a problem with breathing IN but a problem with breathing OUT. During normal inhalation, air moves smoothly from the mouth, through the trachea, bronchi, and bronchioles into the alveoli. If you're an asthmatic, you can do the same: you lower the diaphragm, you swing the ribs out, and that makes the lungs bigger.

However, breathing out is not active but passive. Ordinarily, to breathe out all you do is stop breathing in, and you automatically breathe out. But if you're an asthmatic, you can't do that. The minute you relax your ribs the obstructed airways block the airway flow and air can't get out. You have a lot of dead air space trapped in your lungs, and you end up breathing at the top of your lungs, instead of using the entire lung.

The four components that cause asthmatics to have trouble in breathing are secretion of excess mucous, swelling of the airway, inflammation in the airway, and muscle spasm.

If you don't have asthma and want to get a sense of how, say, your children or parents with asthma feel, try this: breathe in. Now don't let the air out. Hold that deep breath and breathe in and out using only the top of your lungs. It's uncomfortable to breathe up here, and that's what an asthmatic does.

When an asthmatic's airway is full, it contains not only excessive, very sticky mucous but also a lot of other debris, including white blood cells. The airway becomes very irritable and sore. When an asthmatic coughs, or inhales cigarette smoke or irritating fumes or vapors, they still begin to wheeze—a whistling sound—because the airways are very irritable and have constricted.

ASTHMA AND THE OFFICE

Avoiding asthma-causing substances and situations in the home is one thing. Now we need to talk about the same subject at work.

Asthma is a disease that can be triggered by physical and emotional factors. Stress at work is certainly a leading trigger for an asthma attack; but as we all know, stress can occur anywhere and anytime. The usual irritants such as airborne allergens (aeroallergens) and dust are present in most work environments. Chemical smells, tobacco smoke and automobile and diesel exhaust are major triggers of asthma.

What will frequently occur is that too many people will be moved into a room meant for a few. This can happen when a small-to-medium size office is turned into a conference room, lunch room, teaching room, or lounge. Several things happen here. First, in an air conditioned building carbon dioxide builds up and available oxygen decreases. There is just not enough fresh air entering through the supply duct register.

Also, since air is not cycled enough for the increase in workers there will be a build-up of particles that are tracked in from outside as a well as bacteria and viruses, including cold and flu viruses, that come from the people themselves. Similar to the environment on an airplane, diseases will be transmitted in a closed space.

Here it is important for management to take this into account and permit specialists to adjust and balance the air flow necessary to handle the increase in the number of people. This situation not only occurs in the workplace but in hospitals and in schools.

Businesses in many parts of the country rely upon roof-mounted evaporative coolers for their indoor cooling. Whenever these devices are used, musty smells, mold spores and bacteria can be sent indoors from a dirty unit.

How do you suspect that a problem occurs at work?

.You will have improvement during days off, weekends or vacations.

.You can have symptoms in less than an hour in the workplace, within two to eight hours after starting work, and even at night after an exposure at work.

.Symptoms can occur after an obvious exposure, such as a chemical spill or in a closed room with little air movement or when it is too hot or too cold for your comfort.

One example of a worst case scenario at work is when many copiers, computers and other electronic machines are packed into a small room with a lot of persons, as noted above. In ninety-nine percent of the cases the airflow in these rooms is not designed for this dense package of humans and machines. The end result can be an occurrence of asthma, headaches, drowsiness, irritability and other symptoms among persons working in this area.

TYPES OF ASTHMA MEDICATIONS

Cromolyn: This drug helps prevent allergies. It is used in an inhaled form for asthma and in a nasal spray for allergic rhinitis. This drug stops the mast cell from secreting the chemicals that cause allergies. Cromolyn does not work for everybody and it is expensive. But for anyone with allergic asthma, this drug is worth a try and may prevent exercise-induced asthma.

Bronchodilators relax the smooth muscles that line the airways, and open the airways. There are three classes of bronchodilators: anticholinergics, methylxanthines, and beta adrenergic agonists.

Anticholinergic drugs: These stop the action of acetylcholine, which triggers the muscles to contract. They relax the muscle by blocking its contraction. These are medium-strength drugs with a shorter duration of action than adrenergic agonists. They are very important for the treatment of patients who produce excess mucous but are less important for treating asthma. They have been shown to be very effective in some children and adults, especially those with chronic obstructive pulmonary disease.

Methylxanthines: These drugs have become popular in the last 20 years. This class of drug is the major type used in the United States for the treatment of asthma. The most commonly known methylxanthine is theophylline. Theophylline acts by stopping some of the enzymatic actions in smooth muscle cells and thereby relaxing them. This is a very potent form of therapy that can have very serious side effects if blood levels rise too high due to the effects of other drugs, fever, flu shots or other diseases.

Beta adrenergic agonists: These drugs are related to adrenaline, our natural chemical that gives us energy when we are afraid or nervous. Adrenaline makes the airways dilate to enable them to take in more oxygen. So do beta adrenergic agonists. The best part is that these drug are more specific for the lungs and usually do not affe the rest of the body, although some patients get tremo or stimulation from them. Beta adrenergic agonists work within minutes and can prevent exercise-induced asthma.

Corticosteroids: These are probably the most effective drugs for the treatment of asthma. They work by reducing swelling, reducing mucous secretion, reducing numbers of mast cells and secretions, and even by stopping the production of the IgE allergic antibody. However, research suggests that they may decrease attention span in children and produce insomnia in some adults. Recently, there has been some concern that certain drugs or fever can interfere with elimination of this drug from the body and allow levels to rise and become unsafe.

Inhaled steroids have been engineered over the past several years to work in the nose and lungs and have little action in the rest of the body. Hence, they are usually considered to be safe and effective.

Immunotherapy (allergy shots): Useful when asthma has clear-cut allergic causes that are not controlled readily with medication. When immunotherapy works, it reduces not only the need for other treatments but the disease itself.

This information on asthma has been adapted from a lecture presented by Michael A. Kaliner, M.D., as presented in a booklet by the U.S. Department of Health and Human Services, National Institutes of Health. I would also like to thank Dr. Jacob Pinnas of the Allergy Center of Arizona, for his critical review of this information.

ASTHMA TRIGGERS

Allergy is the number one cause of asthma. About ninety percent of the people under ten who have asthma have allergies. If you are younger than thirty, there is a seventy percent chance that you are allergic.

An allergic reaction works in the following manner: Every tissue in the body has "mast" cells. These cells are most heavily concentrated in the mucous membranes (the skin that lines the nose and the airways). When mast cells are sensitized by having IgE (allergy related) antibodies on their surface, they encounter that antigen or foreign substance, then trigger the release of histamine and other chemicals from inside the mast cells, and this causes the allergic reaction.

You don't exhibit an allergic reaction until you have been exposed to the antigen (pollen, mold, dust, etc.) for a period of time. So if you just acquired a cat, just wait!

Ragweed produces about a billion pollen grains per year, per plant. After two or three years (or seasons) of breathing in an allergen like ragweed pollen, the next time that an allergic person breathes in that pollen, instead of simply causing IgE to be produced, the mast cells, now sensitized with IgE, release histamine and other chemicals. Asthmatic reactions now occur. This can occur in an asthmatic who is allergic to certain foods. These same reactions can also occur in response to exercise.

Allergies are hereditary; if you are a parent with allergies, the likelihood is that one-in-three or one-in- four of your children will have allergies. If both parents are allergic, all offspring are likely to be allergic.

When is allergy likely to be a contributing factor to asthma?

. when a blood relative has allergies;

. when the asthma begins at a young age;

. when the asthma symptoms occur or worsen seasonally, such as the fall or spring;

. if other allergic symptoms also occur, such as rhinitis (runny nose), hay fever, or eczema;

. if tests show that blood and sputum contains an increased number of eosinophils.

Infections can also cause asthma. Bronchiolitis is a viral respiratory infection that occurs in children younger than two. A child may get a fairly bad cold and then develop respiratory distress. About fifty percent of these children, if they have an allergic parent, will go on to develop asthma. Generally, this asthma is fairly mild and is greatly improved before age ten.

Some asthmatics only have asthma symptoms in relationship to a cold and sinus infection. You will recognize sinusitis because you will feel mucous dripping down the back of your throat. Often you will have headaches in the sites where the sinuses are, and you may run a fever. Sinusitis commonly causes asthma to worsen. They should both be treated at the same time.

Nighttime asthma can occur when stomach acid backs into the esophagus, or swallowing tube, and irritates its lining. This sets up a reflex action in the chest triggering an asthma attack. This may be due in part to overeating or eating the wrong foods.

Cigarette smoke and other chemical irritants can trigger asthma as can intense emotions and psychological stresses. Triggers also include rapid temperature changes, onset of stormy conditions and cold weather.

In a healthy person, about seventy-five to eighty percent of the air comes out within one second of maximum exhalation, that is, breathing out as hard as possible. By three seconds, the lungs have been essentially emptied. Asthmatics can't breathe in as much because they have all their air trapped in the back of their lungs. Even at six and seven seconds they are still blowing out air. This is one way your doctor can determine if you have the disease.

Industrial and occupational exposure can lead to asthma. Inhaled substances (particulates) can act as allergens or as irritants (chemicals) that do not result in IgE production. The most common cause of occupation-related asthma is the inhalation of substances like toluenes, benzenes, acetones, xylenes, or formaldehyde. Anyone who inhales chemical fumes can develop bronchial irritation and can become sensitive to the chemical. It is estimated that as many as fifteen percent of asthmatics develop asthma in response to industrial or occupational exposure.

A very common cause of asthma is non-steroidal anti-inflammatory drugs such as aspirin. Aspirin is a very potent drug and about five to ten percent of asthmatics will have asthma triggered by aspirin or other aspirin-like compounds. These include phenylbutazone, indomethacin, ibuprofen, and other non-steroidal anti-inflammatory drugs. Aspirin-caused asthma can be a very severe form, and can be prevented by avoiding aspirin and by aggressive treating of the sinuses.

One group of chemical additives that has been found to trigger asthma attacks in susceptible people (about five percent of asthmatics) is sulfites (related to sulfuric acid), commonly found in wines. Sulfiting agents are commonly added to food such as salads to keep them from turning color. The Food and Drug Administration requires labeling of sulfite-containing foods and prohibits salad-bar restaurants from adding sulfites to their fresh foods. However, other processed foods that are available at salad bars may still contain sulfites, as measured in sulfur dioxide (parts per million): instant potatoes at thirty-five to ninety,, pizza dough at eleven to twenty, grape juice at eighty-five, wine at one-hundred-fifty, fruit topping at sixty, lemon juice at eight hundred, instant tea at five to six, beer at ten, dried fruits at two hundred seventy-five, and canned vegetables at five to thirty.

Beta blockers (beta adrenergic antagonists) have been recognized as causing asthma, since their first day of introduction. These chemicals are used for treating of migraine headache, glaucoma, rapid heart rate, high blood pressure as well as tremors and other conditions. You may need to switch to an alternate drug, if you have asthma. Talk to your doctor, since these drugs will make your asthma worse, according to the U.S. Department of Human and Health Services.

Exercise is a potent stimulator of asthma. When a person exercises, they hyperventilate by taking rapid, shallow breaths. Just as evaporating water cools the skin, hyperventilation cools the airways. A reflex reaction to this cooling of the airways causes asthma. Running is the worst exercise for asthmatics due to the expiration of water, and swimming is the best because very moist air is inhaled. This slows down the cooling of the airways. Wearing a surgical mask is helpful at other times to enable a person to re-breathe humidified air. A significant percentage of Olympic athletes have asthma and have made it through years of rigorous training to the highest level of competition by keeping close check on their disease.

Idiopathic (unidentified) causes of asthma accounts for about twenty-five percent of the individuals above the age of thirty. The cause of the asthma is uncertain and occurs most often in older individuals who have some bronchitis, a lot of excess mucous secretion, and perhaps sinus infections.

ASTHMA TREATMENT

There are two major ways to treat asthma: avoidance and medications.

Avoidance: As an example, if you have a pet and you have asthma, chances are you will react to a cat before you react to a dog. The reason? Saliva. Cats clean themselves more than dogs do, and at the recent American Academy of Allergy, Asthma and Immunology meeting, there was much to do about the allergic and asthmatic problems caused by cat saliva. So, if you must have a dog or cat indoors, at least don't let the pet into the bedroom. Even birds should be avoided indoors, since feathers are a potent allergen.

Recognize that airborne pollen levels are highest in the early morning hours on bright, sunny, and windy days. Use air conditioning in your car during pollen season whenever possible.

One myth about asthma is that children will outgrow it. In fact, this occurs in only about fifty percent of children. It often recurs when they reach their thirties and many others still have abnormal airway disease. But there is no reason to let a child suffer, waiting for the chance that he or she might get better. Have your child seen by a specialist in allergy or asthma. Eight percent of the athletes who represented the United States in the 1988 Olympic Games had exercise-induced asthma. Proper medication helped them come closer to their full potential.

This series on asthma has been adapted from a lecture presented by Michael A. Kaliner, M.D., as presented in a booklet by the U.S. Department of Health and Human Services, National Institutes of Health. I would also like to thank Dr. Jacob Pinnas of the Allergy and Asthma Center of Arizona for his contributions to this series.

CHAPTER 3
A HAZARDOUS WORLD

HOUSEHOLD DUST

The vast majority of indoor allergies are caused by house dust. Just what is this stuff that causes perennial allergic rhinitis, bronchial asthma and respiratory allergy? People with house dust allergy usually feel worse when indoors, during the night, and year round and generally feel better when out of doors.

House dust is a very complex mixture composed of fragments and feces from insects such as moths, cockroaches, ants, silverfish, spiders and mites. It is composed of plant hairs and flower parts that are tracked indoors; pollens and spores; fibers of material made from cellulose such as cotton, wool, linen, jute, wood, kapok; man-made fibers such as fiberglass, nylon, plastic, dacron and rubber and a dozen others; animal and human hair and skin cells; cigarette smoke, fireplace soot and diesel exhaust; lead, insecticides, aerosols from personal care products and cat and dog antigen. It includes food particles as well as tracked-in outdoor dirt that becomes worn to a powder.

One study has shown that the average six room home can accumulate up to forty pounds of dust per year. If you are curious, you can sift through your carpet cleaner bag to see the larger particles of what is breathed. This settles on the heater to get burned in the winter.

The big problem children in this house dust family, however, are mite feces, cat antigen, cockroach parts, animal dander and mold spores. The frequency and amount of these depends on your living conditions and location.

Humans shed about one gram of skin cells per day. Mites survive on this dander as well as on the shed dander of animals.

The chief problem with house dust comes from the ninety percent that is settled, not the ten percent that is in the air. The worst areas of the home for house dust are the carpet (especially deep pile), upholstered furniture, mattresses, box springs, blankets, bedspreads, comforters, quilts and drapes, as well as stuffed animals. Old upholstered furniture is especially troublesome. Usually the older the furniture, the more allergy it causes, because of a deeper layer of dust and a large reservoir of dust mites.

Not too many decades ago people were concerned about dust in the home and decided that it was better to keep it out since, for whatever reason, it aggravated allergies and asthma. As technology improved we started looking for different things in dust and found them—some half a hundred or so, serious things such as lead, pesticides and benzene. Over time technology allowed us to put numbers to what we found and we could tell what was "normal", "low" or "high", whatever any of those terms meant.

The next thing that happened was that we were able to detect many of these substances in the human body, see how much of it was there, run tests on the humans as well as conduct surveys, and determine what these things were actually doing to us. In a word we started to make some discoveries and got a few surprises.

Now we are starting to look at combination effects; where exposure to two pesticides gives a thousand-fold change in tissues rather than a one plus one change.

It is no secret that the rate of cancer and various respiratory diseases has been increasing in the United States. The problem has been that a hundred billion dollars have been going toward outdoor air pollution over the past few decades while we spend some ninety percent or more of our time indoors. Our exposure to toxic substances indoors is some five to ten times greater than our exposure at the outdoor level. In some cases it is one-hundred-fold higher. This includes pesticides, lead, heavy metals and cancer causing volatile compounds. Most of the problem is associated with two types of hazards: those that start out in the air and end up in house dust, and those that start in outdoor dirt and end up in house dust.

These investigations are reported in Scientific American, Review of Environmental Contamination and Toxicology, Clinical and Experimental Allergy and a score of other scientific journals.

These journals tell us exactly what is known about a subject area at the time they are written. The information may be changed or disputed as more information comes to us but at the time, it is the best that we've got.

Indoor exposures and reactions to toxic agents, particles as well as gases far exceed those limits set by the EPA for work, basic outdoor or Superfund sites (toxic waste dumps). This is because governmental agencies never thought that indoor air quality and indoor dust was a real problem. Fortunately, the picture is changing.

SOFT FURNISHINGS

Soft furnishings retain a high level of dust as well as the highly allergenic cat antigen. Take a sofa pillow outside, stand upwind and beat it to see what I mean. Every time a soft furnishing is sat upon or even touched dust is released into your breathing space.

If you are in the market for furniture, look for solid unwoven or tight-weave fabrics. They may be cotton or synthetic. These will retain very little dust. Be sure to ask about the amount of formaldehyde that is in the wood frame of the chair or sofa as well as the amount that is in the fabric itself. The current scientific thought is that formaldehyde is an irritant only, unless you are chemically sensitive. For most persons it has an odor that goes away in a short period of time.

You may want to shop for an appropriate cover for that sofa and chair. This is more economical than buying a new one. And you might as well think about removing furniture that is unused. It is only a dust collector and it still has to be moved to clean underneath it.

Vacuuming can raise dust levels as much as 1000-fold, if improperly performed. One study actually found that it was better not to vacuum than to vacuum, as far as keeping down the level of airborne dust. If you remove the carpets then you have the choice of damp mopping or vacuuming. On the other hand, carpeting adds visual warmth and important insulation to many homes. You will have to decide. If you have a wood floor underneath the carpeting get some larger non-slip throw rugs or center pieces instead.

Household dust settles near the walls because of its tendency to swirl outwardly as we walk; so concentrate your damp dusting (or vacuuming) at the edges.

The bed can be covered with a nice looking, tight-weave blanket to catch the dust and which can be carefully folded back at night. If they aren't too large, bed spreads can be thrown in the dryer on "Air" cycle to remove the dust. Otherwise, they may have to be taken outside, weather permitting.

Unfortunately, the presence of snow and ice does not cut down on indoor dust since it is all tracked in now, with less blown in. Since tracking is the primary entry of dust into the home leaving the shoes at the door becomes important.

Blinds should be vertical, not horizontal; curtains and drapes and lamp shades should be smooth and solid, not porous or rough.

Deep pile carpets catch and hold a tremendous amount of dust and it is almost impossible to properly clean and vacuum them. The best bet is to have that deep pile removed.

Minimize the soft furnishings and minimize the dust and the allergens. Cut the number of items and cut the amount of dust and housework.

OZONE

First some basics. The air concentration of ozone at ground level is between 10-25 parts per billion (ppb). Smoggy cities will have ten times this amount. It is considered to be toxic at 60 ppb. Children are more sensitive than adults to ozone and will have decreased lung function at the 60 ppb level. Most manufacturers of ozone generators recommend that they be used in the 30 to 50 ppb range. The response to the odor can vary greatly from person to person.

The issues:

Does ozone remove odors, and if so what kind of odors does it remove?

Why do some people swear by the use of ozone to rid their homes of problems?

Does ozone reduce the incidence of allergy and asthma within a home or building?

Why is ozone considered to be unsafe and at what level is it unsafe?

Ozone is promoted as a biocide; that is, it will kill mold and bacteria? Will it do this?

How does ozone treatment compare with an air purifier for removal of respiratory irritants?

According to the World Health Organization there are some three hundred chemicals that can be identified in the air of the average home. Both laboratory and in-home experiments have found that ozone reacts with only a certain small class of these chemicals. What is created from the reaction, is not pure carbon dioxide and water, as dealers and manufacturers claim, but formaldehyde! The level of formaldehyde can actually increase with ozone treatment.

Ozone was not found to react with or decrease the amount of formaldehyde itself.

When reacting with volatile organic compounds (VOCs) ozone produced a number of new compounds. Total VOCs increased in the home in the presence of ozone.

In addition to the production of formaldehyde the chemicals that ozone does react with may produce odorless free radicals that may be more hazardous that the original odor causing agents.

Do odors decrease with the use of ozone? Yes, say researchers. The reason is three-fold. First, it is believed the odors decrease, and so they do. Second, ozone is a classical masking agent and the pungent odor of ozone covers the smell of odors. Third, many researchers have found that ozone affects the sense receptors in the nose which inhibits our ability to smell. This is why people swear by the use of ozone, because they can't smell the odor anymore.

What kind of odors doesn't ozone work against? According to published reports these include tobacco smoke, detergents, waxes, cleaners and scented room fresheners. In addition to these the three primary off-gases from carpeting are converted to formaldehyde. Even at a high level it would take between 880 and 4400 years to remove household odors with the use of ozone, according to data presented in a recently published industrial hygiene journal.

How do we know these things? Because we can monitor the exact chemicals with which we started in a home, and find out with instruments that they are still there after ozone treatment. This is even when people say they can't smell the odors any longer.

Does ozone treatment reduce the incidence of allergy and asthma? Research suggests that low ozone concentrations can increase sensitivity of the airways to allergens and aggravate asthma. One school of thought is that this is one of the reasons why respiratory disorders are on the rise in big cities.

Experiments using lower and lower concentrations of ozone still report damage to the respiratory tract, often permanent. Studies have been conducted on animals and people and conclude that the EPA needs to lower its standard of what is an acceptable amount of ozone in our environment.

Some manufacturer's of ozone machines are in trouble with Federal Trade Commission (FTC) for making false claims of health; the FTC doesn't recognize ozone treatment as a legitimate air cleaning method; ozone supporters are in trouble with the American Medical Association and the American Lung Association for putting people at risk; in trouble with the American Society of Heating, Refrigeration and Air conditioning Engineers (ASHRAE) for the same reasons.

The policy statement of ASHRAE is that while ozone can oxidize odors in water the level required to do this in air would be harmful to building occupants. "The major effect of ozone generators is to reduce sensitivity of the sense of smell, rather than reduce actual odor concentrations."

Why is ozone under attack from virtually all quarters? The ozone people do not come up with hard data that is contrary to published scientific findings. Enthusiastic ozone supporters make false claims that get them in trouble. Ozone manufacturer's are overselling a product that is proven by comparison tests to be second best to a basic home air purifier with a filter-activated carbon combination.

Ozone has been used for years as post-fire odor control. The level of ozone used for this purpose is extremely high and does not apply to the low concentrations that are promoted for in-home usage.

While active research on this subject has been going on for the better part of a century, home ozone devices are selling faster than ever. Devices are available for the central air system as well.

If ozone is supposed to be of such great benefit, why is it so hard to prove?

SLEEPING WITH THE WINDOW OPEN

Many allergic persons will sleep with the window open at night for fresh air. Their thinking is that since pollen is released by most trees and many weeds in the morning hours, or in the afternoon at latest, that there will be few respiratory hazards to breathing cooler and fresher nighttime air.

What really happens is that the cooler and denser air mass in the dark morning hours slides down the mountain slopes and contours of the ground to settle into locations of low elevation. The locations can include generally low lying areas within the city itself or even in your neighborhood. Anybody who has ridden on a motorcycle and driven down into a dip in the road will know what I mean by cool air masses in low lying areas.

These air masses can be a thousand feet in height or much less, but they can bring down with them aeroallergens that have built up in the atmosphere during the daytime hours. A heavy onset of outdoor dust can do the same thing as the dust particles capture the pollen, mold, and a variety of particulate pollutants to cause them to settle out at high concentration.

DUCT CLEANING

Duct cleaning is over a billion dollars big in the U.S. The National Association of Duct Cleaners even has its own congressional lobby. There is enough business in many communities for a dozen duct cleaning companies, although some would say that there are too many. Many owners of duct cleaning companies have engineering degrees and have a good understanding of the entire air handling system.

Each home is different in terms of the age and general health of the furnace, air conditioner, evaporative cooler and ducts. Homes are of different sizes and shapes. Some homes are located in dustier or more humid locations than others and need more frequent cleaning, not just within the home proper but the ducts as well. Duct cleaning is not the cure-all, however it does have its place in terms of overall health of the home.

In a high percentage of homes, air ducts have a serious leakage problem. They will pull in particulates, gases and allergens from the attic or from the out-of-doors. When ducts leak, heating and cooling in the home is uneven. This can result in microbial contamination of the ducts and various areas of the home and increase your power usage. If you are going to have your ducts cleaned you should also ask about having them pressure checked for leaks.

Who needs to have the ducts cleaned? New homes with residual construction material, older homes that have not had their ducts cleaned, homes that have had a small fire or chemical spill (including extensive painting), rusty furnace, or if you have gone a few years without a cleaning. Everyone should have their ducts pressure checked for leaks since leaky ducts are extremely common, whether the home is new or old.

What is the average amount of dust removed? In terms of three-pound coffee cans, somewhere between one-half to fifteen of them with the average around two or three.

The registers should be disinfected. Evaporative coolers deposit more grit in the ducts compared with air conditioning systems. This is understandable. They also contribute moisture and in a dirty duct system microbial build-up can result.

What should the consumer be concerned with when looking for a reputable duct cleaning company? First, find out if they are licensed, bonded and insured. This allows the consumer some recourse if the business does not live up to its promises. And as always, shop around and compare.

Recently, researchers have published what may be the only report in the scientific literature regarding sanitation of the central air handling system, or "air duct cleaning." The results were favorable. (Note: NADCA, or the North American Duct Cleaners Association, recommends against the addition of any agent to the air duct system, unless extreme circumstances exist.)

This investigation took place in Fort Worth, Texas. Six houses were studied in the winter and six in the summer. Two homes from each group served as untreated controls for comparison. The homes were all air conditioned and ranged from five to sixty years of age with an average home age of twenty-seven years. They were constructed of brick or brick and wood and all homes had light to heavy fungal (mold) growth on the air conditioner fan, coils, drip pan, supply air plenum or registers.

Excess moisture leads to mold contamination in central air handling systems. This occurs due to improperly installed humidifiers, an AC system too small for the cubic footage that it has to serve, dirty cooling coils or poor moisture drainage from the drip pan, usually due to blockage. The homes were sampled for mold at the supply or return air vent.

After the homes were initially sampled the experimental homes were treated in the following manner: the vent registers were removed, taken outdoors, and cleaned with a 0.25% glutaraldehyde solution.

Each outlet, the return air ducts, and the air conditioning/heating units were vacuumed with a HEPA vacuum. The cooling coils were disassembled, cleaned and reassembled. The ducts were fogged with an antifungal agent and the vent openings in each room were also cleaned to remove dust and spores. A permanent, washable electrostatic air filter was installed in each system. Homes were tested over a two month period.

Eight weeks after sanitation, the study houses demonstrated an overall reduction in mold of ninety-two percent during winter and eighty-four percent during summer. The control homes demonstrated no reduction in mold over this same time period.

The lessons here are that the AC should be inspected and cleaned periodically and a filter of good quality should be installed in the central air handling unit. While the actual ducts were not cleaned in this investigation it is reasonable to assume that this action, coupled with the above measures, should result in a further reduction of mold contamination, especially in older homes.

AIR HANDLING SYSTEMS

In a 1989 report to congress, the EPA stated that "sufficient evidence exists to conclude that indoor air pollution represents a major portion of the public's exposure to air pollution and may pose serious acute and chronic health risks." The focus here is on office buildings, hospitals and schools. There is a lot of truth here that applies to homes as well.

Given that heating, ventilation and air conditioning (HVAC) equipment is in place, most professionals will agree that problems are rarely with the equipment itself. Usually problems will occur with the installation or upkeep of the machinery.

When air quality problems in buildings arise then the finger points at overloaded equipment and misused airspace. There are many examples. One example is not maintaining clean air conditioning drip pans. This would permit the build-up of mold and bacteria. Other examples include: the infrequent changing of HVAC filters, or using the wrong filters for the system, or placing fresh air intakes downwind from exhaust vents, or placing ten persons in a space meant for two. This applies to schools as well as to commercial buildings.

Remember that air conditioning does not remove smells, it just spreads them around, or dilutes it, in the case of commercial systems. Frequently the renovation process overloads the indoor air with the smells of adhesives, new carpeting and paint without adequate warning to building inhabitants. Almost all of us have experienced this gross lack of concern for building occupants by building management at one time or another.

Activated carbon filters are now available for the home and office building that will remove indoor odors without impeding air flow. These are available through air filtration companies commonly listed in city telephone directories.

It is the responsibility of building management to advise building inhabitants to use common sense ventilation. This means that those who are leasing the building as well as the management both share in the responsibility of maintaining clean air. And just because the air filtration system may be state-of-the-art does not give us a license to be careless. One cannot buy a car with airbags and then drive recklessly and still expect to be safe.

Many experts advocate a prohibition of the introduction of chemicals into the HVAC system. This includes the ducts, except as may be required in an unusual or emergency situation. Then complete information should be provided to building occupants before any such action is undertaken. This avoids unnecessary health risks, complaints and lawsuits.

There should not be introduction of such additives as deodorizers, disinfectants, perfumes, scents, ozone and other aromas into the HVAC system. Such additions encourage and contribute to the development of sick buildings and unhappy occupants.

Pine scented products and most particularly air fresheners used to mask troubling odors can be debilitating exposures for those already sick from indoor pollutants. They can be annoying and interfere with worker efficiency and personal vitality. We should strive to appreciate air without smell and should encourage the use of self-operating, ratcheted or sliding, transparent and silicon derived ventilation systems, also known as windows.

If you or your family members have asthma or perennial allergy you should think seriously about converting your home to air conditioning. This will eliminate most of the potential problems associated with the evaporative cooler.

THE FIREPLACE - I

Changing pollution standards around the country are causing a phase-out of wood burning fireplaces. If you still use one then you need to know how to avoid breathing the toxic wood gasses and prevent them from absorbing into your soft furnishings.

Several factors can lead to entry of fireplace smoke into the home:

If the air is very cold outside and your fire is slow to start then there will be insufficient heat to break through the plug of cold air in your chimney. Smoke will back into the home. Remedy: Prepare your burning material but don't light it. Then hold a wad of lighted paper up into the chimney throat to warm the air there and then light the fire.

Many people will go to bed with the fireplace going strong. If the fireplace and the home cool off sufficiently in the early morning hours, this will cause a suction of air down the fireplace and the smoldering fire will enter the home. You may want to consider setting a timer or alarm to permit you to get up and close the flue, sometime after you have gone to bed.

In tight newer homes, if the forced air furnace goes on while the fire is going then air from the home can go into the furnace. This results in negative inside pressure. Likewise, kitchen and bathroom fans that are on or a clothes dryer vent that is in operation can have the same effect.

Downdraft can also occur for the following reasons: The chimney is too short. Or trees, hills or other buildings nearby can deflect the wind downwards; older chimneys may not meet code; burning too large a fire before the chimney has had a chance to warm up; and not opening a window on the windward side of the home.

It is relatively inexpensive to install an outside air kit to your fireplace. This will ensure that the wood burns efficiently. It will also avoid the necessity of opening a window for a supply of air to support the fire. This could pass a cold draft across your neck.

Carcinogenic creosote is always given off during wood burning. It is also flammable when it accumulates in the chimneys of older homes. It is more likely to accumulate if too much wood is burned too fast and combustion is incomplete. Plastic will also stick to the inside of your fireplace, when burned. Also, never burn foil, painted wood or garbage in the fireplace since, along with plastic, they will produce poisonous fumes.

As far as the wood itself, soft woods such as fir and pine are easy to light and burn rapidly with a hot flame. Hardwoods, such as ash, beech, birch, maple and oak provide a longer-lasting fire with a shorter flame.

Ideally, the wood should be cut in the spring so that it is dry by winter. This way the moisture content will be less than twenty percent and it will burn more completely. Heat energy will not have to be lost vaporizing the water. Dry seasoned wood will have cracks on the end extending from the center like bicycle spokes. The bark is easily peeled with no green showing. Finally, it should be light and not sound like a thud when hit together.

Always have an approved fire extinguisher handy, whether you have a fireplace, wood burning stove or neither.

If you have a gas line stub to your fireplace then you can use special gas logs and avoid all the problems of wood burning. These include uneven heating, greater fire danger, outdoor air pollution, as well as the necessity to clean out the ash, and pollution. Gas is much more efficient as a fuel (heat given off per dollar spent), rating at about thirty to thirty-five percent efficiency compared with five to ten percent for wood.

But it is unvented gas appliances that are the industry's fastest growing product category and its most controversial. You decide.

Unvented gas heaters and fireplaces appear to be a good industry in which to buy stock. Driven by almost unbelievable consumer demand they are touted as being 99.9 percent efficient. One manufacturer's products are capable of burning permanent gas logs at an altitude of up to 10,000 feet.

It appears that they become less efficient when there is poor ventilation such as in an airtight newer home. Less efficiency means the production of increased amounts of poisonous carbon monoxide gas. On the good side there have been no documented deaths attributed to unvented gas heaters and fireplaces.

Many dealers have drawn the line, however, and do not want to be part of a potentially dangerous product. In addition, there are still issues to be addressed regarding the production of carbon dioxide and nitrogen oxides by the heaters. There is also a problem with a serious reduction of indoor moisture.

Kerosene heaters were developed by the Japanese in the 1950s and soon were used by the millions in Asia. By 1983 there were twelve million in use in this country. They were ninety-five percent efficient but misuse, bad press, and poor marketing practices sealed their decline into oblivion. The Consumer Products Safety Commission investigated the matter and determined that the units themselves were not a problem. Problems were related to human error: using the wrong fuel or placing the unit too close to combustibles. But it was too late.

The pellet stove is an exciting new heating technology. The stove burns waste wood that has been compressed into a pellet about the size of rabbit food. The waste sawdust from sawmills provides the wood for the pellets. A gravity feed drops the pellets into a closed combustion chamber. A fan provides air circulation and the burned gases are vented outside. Very high temperatures are reached with this method of burning since the pellets are dry and combustion is virtually complete. It has been found to be the cleanest wood burning technology to date.

Pellet stoves can be freestanding or inserted into fireplaces. Mobile home units are available. The production of toxic gasses in the home is not an issue with pellet burning devices.

THE FIREPLACE - II

Are you starting to think about the fireplace this season? One by one all of our old traditions fall by the wayside. So it is with a good old fashioned fireplace and the wood that it burns. But maybe you don't have to stop using it for health reasons. Just maybe you can fix it right so that it burns cleanly and safely.

First, the worst case scenario. Given that wood smoke is not breathed in immediately it can reside in soft furnishings for weeks and months. The smoke has a lot of components. Let's look at a few of them.

.PM-10: These are very small particles less than 10 microns in size. They are smaller than pollen grains and can deposit more deeply in the lung where some of them can remain indefinitely to cause damage. Rapid declines in lung function can occur upon exposures to PM-10s. Breathing problems can persist for up to two to three weeks, according to several studies. The gases mentioned below can attach to the particles to compound the problem.

.CARBON MONOXIDE (CO): CO combines with hemoglobin, the oxygen carrying substance in the blood, replacing the oxygen. The result is a greater incidence of angina among persons with cardiac disease. CO also affects the chemically sensitive persons and "normal" persons to result in lethargy, drowsiness and headaches.

.ALDEHYDES (including formaldehyde and acrolein): Exposure to formaldehyde above 0.4 parts per million is associated with upper airway irritation, headaches, and disorders of the nervous system. Also, it is presumed to be a carcinogenic risk to humans. Acrolein, another aldehyde present in wood smoke, is an even more active irritant of the eyes and respiratory tract.

.NITROGEN OXIDES (Nox): NOx is known to cause a build-up of fluid in the lung and causes ring in the lung at high concentrations. Children from homes with gas cooking stoves (which emit NOx) have a greater frequency of respiratory illness than do children from homes with electric stoves.

.POLYCYCLIC AROMATIC HYDROCARBONS (PAH): Many of these compounds are carcinogenic in animals. They also include chemicals that result from the burning of scented trees such as juniper, cedar, pine and eucalyptus.

All of the above chemicals are found inside homes that burn fireplace wood. Obviously, the burning of plastics, textiles, laminated wood and other materials can worsen problems.

What really happens is that the cooler and denser air mass in the dark morning hours slides down the mountain slopes and contours of the ground to settle into locations of low elevation. The locations can include generally low lying areas within the city itself or even in your neighborhood. Anybody who has ridden on a motorcycle and driven down into a dip in the road will know what I mean by cool air masses in low lying areas.

These air masses can be a thousand feet in height or much less, but they can bring down with them aeroallergens that have built up in the atmosphere during the daytime hours. A heavy onset of outdoor dust can do the same thing as the dust particles capture the pollen, mold, and a variety of particulate pollutants to cause them to settle out at high concentration.

HOUSE FIRES

Expect smoke from a fire to go anywhere that air goes. The soot and the odor are toxic to different degrees in different people, or the same person at different times. This is true of any respiratory irritant. Good restoration of the home after a fire means removal of soot and odors. Restoration specialists know how to do this.

Most people are shocked at how much and how long their lives are disrupted after a fire occurs. If even a small fire occurs in your home expect to do some or all of the following: have the carpets, window coverings, furnishings and clothing cleaned; launder bedding and personal clothing; walls and ceiling repainted or replaced; oil paintings and frames need to be restored; windows repaired; plastics discarded; food and cosmetics replaced. The furnishings may have to be removed and cleaned inch by inch with a cotton swab.

The ducts will also have to be cleaned., otherwise soot could be released into the home when the system is turned on.

A non-toxic non-allergenic chemical deodorizer is sprayed in the home to eliminate as much smell as possible. This takes two-three hours. A high concentration of ozone treatment of the home will do the same thing but will take eight hours or longer. You will have to be gone from the home in either case. In short everything will have to be cleaned or replaced.

You will need to start a special file listing items that were destroyed, cleaned or replaced for insurance purposes. The entire matter is a lengthy and tedious process. Unless the restoration process is done right the smell and the soot will remain and could affect the health of your family and household guests. It could also lessen the resale value of your home.

Black soot is not always a marker of smell. For example, chicken left forgotten for a couple of hours in a teflon frying pan will give off the foulest odor you can imagine. Much of the home is covered then by a white film that is carcinogenic. The carcinogen is related to pesticides that contain estrogen-like compounds.

It is important to note that fire from synthetic materials is more likely to contain more dangerous gases than when natural materials are burned. Synthetics include most carpets, chairs, sofas, beds and wall paneling. This is especially true when plastics and materials from hydrocarbons are burned. It includes products with formaldehyde, styrofoam, nylon and polyvinylchloride (PVC). The gases released include phosgene, cyanide and hydrochloric acid, as well as thousands of others.

The presence of smoke from a fire can not only impair vision but can lead to loss of consciousness and irritation of the entire respiratory tract. The results lead to physical weakness, loss of coordination, faulty judgment, blurred vision and panic. Survivors of a fire may also experience lung problems after the event.

In addition to fire gases, smoke also consists of fine particles and suspended liquid droplets known as aerosols. Smoke particles can be harmful when inhaled and exposure over too long a time can cause damage to the respiratory system since the particles are small enough to be inhaled deeply. These fine smoke particles have toxic gas molecules attached to them which makes them particularly hazardous to inhale. This is because of their tremendous surface area to volume ratio.

Small household fires need to be treated seriously from the standpoint of watching the health of the occupants. Asthmatic patients and others with a wide variety of respiratory disorders suffer immediate and long-term effects from fire smoke. Long-term exposure and harm can occur because toxic gases and particles are absorbed into the soft furnishings of the home. The drapes, beds, clothing, carpets and curtains will absorb the most smoke followed by the walls and ceilings. These substances will emit low to moderate concentrations of toxic gases for weeks, months or years later. The residents and visitors to the household will experience this exposure during this time period, through inhalation as well as skin absorption.

If a fire has occurred in your residence it is important to ensure your health by consulting with businesses that specialize in fire cleanup and detoxification. Frequently the home and furnishings are treated with a very high concentration of ozone which neutralizes many of the harmful chemicals but your furniture may have to be moved out to be treated separately. If you are having symptoms as a result of the fire it is important to consult with your doctor.

CANDLES: SOOT AND AROMA

Carbon soot can come from a variety of combustion sources in addition to candles. These sources include fireplaces, water heaters, furnaces, standing pilot lights, cigarette, cigar and pipe smoke, cooking byproducts, gas dryers, automobile exhaust and fires in general.

One reason there is a rise in the soot level in home is due to an increasing trend in candle purchases. Believe it or not candle technology, or the lack of it, is partly to blame for this. First, manufacturers tend to increase the amount of aroma in a candle. A common result is this is incomplete combustion and a rise in both soot and volatile organic compounds (VOCs). Also, amateur candle makers can produce inferior quality candles for the home with the same results.

With a candle there is a tradeoff—brightness for soot. The brightness of a candle is due to the volatility of the wax and the yellow-white part of the flame. This color of flame is relatively cooler than the hotter blue part of the flame. The soot is caused by incomplete combustion of the wax as it burns in the cooler part of the flame. The bigger the wick, the bigger the flame, the brighter the candle, then the more soot that is released.

Wicks that curl as they burn keep the wick short and hot but not very bright. Shaking the candle or a light breeze can cause the candle to release soot.

Particle sizes sometimes offer a clue as to their origin. For example, most homes with garages have carbon soot in the air of the home. Most soot particles are in the kitchen or areas nearest the garage entry. A wide variety of soot sizes can be found here (microscopic). The larger particles settle out as the air circulates through the home and one is left with the very smallest particles in the master bedroom, given that it is distant from the garage.

Carbon soot from candles will be present throughout the home but mostly in the room in which the candle is burned.

Many of those who support the use of burning herbs for treatment of asthma recommend the burning of cajeput, cypress, eucalyptus, frankincense, hyssop, lavender, lime, myrrh, spearmint and spruce. Different types of therapy have revolved around using these plants for thousands of years.

Aromatherapy is not healthy for persons with allergies and lung disorders. After all, how could a candle made from ragweed improve the quality of your life.

In point of fact, any fragrance candle can trigger an asthma attack.

BURNING USED TIRES

The issue of what to do with used tires has become a world wide environmental problem since only a fraction of them can be used for asphalt or construction and they are a problem in landfills since they tend to "rise."

The EPA lists one-hundred-twelve companies in the United States that burn tires as a sole fuel or as a fuel supplement to coal, coke, fuel oil or natural gas. These plants burn tires in the range of 200 to 300 tires per hour.

Generally, tires contain the same level of heavy metals as coal. The exceptions here are chromium and zinc. Chromium is present in steel belted tires and zinc is necessary for the rubber annealing process. Chlorine is also present in tires in higher concentrations than in other fuels.

When tires are added to another fuel source the general trend is for the amount of carbon monoxide (CO) emissions to increase due to incomplete combustion. Along with this is the emission of dioxins and furans, the most toxic carcinogens known. Dioxins and furans are formed as a result of incomplete combustion in the presence of chlorine. The emissions of dioxans and furans have not been monitored adequately in many cases of tire burning. The issue of CO emissions is also arguable due to special interests that want to burn the more economical tires, variable presentation of data and variable amount of tires that are burned in differing coal or tire burning plants.

The important point here is that a tremendous number of tires is taken out of the environment and burned for energy. The trade off is increased atmospheric pollution that is well above state standards in many cases.

There are a number of possible solutions including the maintenance of tight state regulations and closer regulatory scrutiny regarding emissions limits for carbon monoxide and carcinogens.

ROOF AND STREET TARRING

Hazardous chemicals are released into the air by the process of street or roof tarring. These can find their way into the indoor air through cracks in the window and door frames or through open doors and windows.

These agents come in two general categories: vapors and particulates. The vapors primarily consist of volatile organic compounds (VOCs) such as benzene, toluene, sulfur, aldehydes (especially formaldehyde), sulfur dioxide, carbon monoxide, naptha and kerosene. The list reads like a Who's Who of toxic gases and for the chemically sensitive person there is no guarantee that a healthy whiff will cure what ails you.

The particulates are classed as polycyclic organic matter (POM) and are tiny globules that range in size from less than that of a bacterium (one micron) to about the size of a pollen grain (twenty-five microns and larger). Under the microscope they appear black and spherical and consist of the solid state of the above gasses. Time-release capsules.

Street and roof tarring is necessary in today's society. But what is the tarring process? Street tarring is performed using "cutback asphalts" which fall into the categories of rapid cure, medium cure and slow cure road oils depending on the circumstance of use and the area of the country. Asphalt cutbacks are prepared by blending asphalt cement with heavy oils, kerosene-type solvents, or naptha and gasoline solvents. These liquids are added to the oil to various extents, depending on how thick one desires the mixture.

No control devices are employed to reduce evaporative emissions from cutback asphalts. In many cases "emulsified asphalts" are used more and more frequently. This involves cutting the oil with detergents which are added to water, rather than with the use of toxic solvents. Limited test data suggest that from RC asphalt, seventy-five percent of the solvent is lost on the first day after application, ninety percent within the first month and ninety-five percent in three to four months.

Evaporation takes place more slowly from medium cure asphalts, with roughly twenty percent of the solvent being emitted during the first day, fifty percent during the first week, and seventy percent after three to four months. The evaporation rate will be much slower in cooler weather and more rapid in the case of roof tarring where less material is involved. And the odors can carry for several blocks.

Armed with this knowledge you might consider renting a motel room for a day or two if tarring is scheduled in your neighborhood.

In a nutshell, tarring in your neighborhood can be annoying at best. The VOCs will be emitted for months after application. This is an important consideration when shopping for a new residence when health is of concern. Find out if the roads are scheduled to be tarred or chip sealed in the near future. If you are chemically sensitive this is one more factor in your decision making process regarding your move to this area of town.

WHAT IS RADON?

Radon is a colorless odorless gas that is a byproduct of uranium. Since it is a gas, radon can attach to dust particles and can be inhaled in this manner as well.

Some say it causes cancer and some say it is not a problem. It's still open to question. Radon is inhaled and is suspect in the cause of lung cancer if you are exposed over a period of decades. To date no cases of lung cancer have been directly linked with radon found in homes. The EPA says that there is no safe level of radon. This does not agree with their own statement that no cases of lung cancer have been directly linked with radon found in homes.

Can we believe all this? I have my doubts.

The whole radon scare may be just a numbers game with the EPA having to justify what may cost consumers twenty-five billion dollars to have millions of homes tested and repaired. And it may be meaningless. Just like spending billions to have asbestos removed from the nation's schools, when asbestos is not a problem in schools. But that's another story.

Uranium deposits can be found in almost any community, to a greater or lesser extent. It is most common in granite, shale and limestone. Uranium has a half-life of about 4.5 billion years. It decays into radium with a half-life of 1620 years. Radium, in turn, decays into radon with a half-life of less than four days. Uranium miners have been found to have a greater than average incidence of lung cancer, separate from smoking.

By the year of 2000, the EPA wanted forty percent of all residences tested for radon gas. As of that date, only about seven percent of homes had been tested. The EPA recommends that every home be tested because of hot-spots wherever you may live.

The Journal of the National Cancer Institute reports that they could find no link between household levels of radon and lung cancer. Conclusion: "... no important public health impact for indoor radon exposure."

A University of Pittsburgh radiation physicist analyzed data from 400 U.S. counties. He found that those with the most radon had the least cancer! Many studies have supported these findings.

Short-term tests for radon can be carried out over two days and can result in more or less than four picoCuries per liter. This is the federal standard. In these tests the home is closed as much as possible over the test period. These short-term tests are not very accurate but are very popular because they give a quick answer.

More accurate testing should be performed over three to twelve months. Two canisters should be used in all testing cases. In these long-term tests the home need not be closed and normal living conditions can be employed. Testing over the long-term frequently gives much lower readings.

CHAPTER 4
WE'RE COVERED IN CHEMICALS

A little fragrance is all right but a lot is all wrong.

Historically, the ancient Greeks used perfumed headrests for the treatment of insomnia. The Romans went to extremes and burned feathers and leaves to stimulate respiration. In today's world the average man or woman is at the bottom of a vast marketing system. As a result we are covered literally from head to foot with untested or poorly tested fragrances, chemicals, sprays, body lotions and preservatives.

A popular shampoo for women has twenty-two chemicals plus fragrance. Another brand has as many as forty ingredients. The male version has somewhat less than this number plus fragrance. But most men have less hair. Hair conditioner for men may have twelve ingredients plus fragrance.

The typical bar soap will have as many as 18 ingredients plus fragrance. The fragrances in bar soap and shampoo become volatile (airborne) to be inhaled in the shower. The effect is enhanced by the chloroform and smell of chlorine that comes from the warm water tap.

Our heads and bodies are dried with towels washed in laundry detergent with fragrance with a little chlorine bleach thrown in and placed in the clothes dryer with an anti-static product plus fragrance.

Skin care lotion has twenty-four ingredients plus perfume. A popular self action tanning lotion has thirty-four ingredients.

After showering most people will use deodorant with about ten ingredients plus fragrance and possibly a talc with fragrance.

Hair spray has eighteen ingredients plus a propellant and fragrance. Not only that but it has acrylate copolymer thickening agent and a form of plastic that is dissolved in an organic solvent.

Many people will use body lotion or body talc after the shower. Men will use after shave. Virtually all of these are scented. Make-up frequently contains formaldehyde.

Toothpaste may contain ten ingredients plus flavor enhancer; mouthwash can have twelve ingredients plus "flavor."

It's easy to see how our lungs, hair, skin and clothing come in contact with an ocean of solvents, chelating agents, emulsifiers, dyes, perfumes and enhancers.

We are exposed to as many as a hundred or more chemicals and a half-dozen fragrances and we're not even out of the bathroom yet!

Are we involved in something that is bigger than we are? Are we losing the war for better respiratory health because somebody else is telling us what is good for us? Are we blaming the pollen for our breathing problems when the answer is in our bathroom?

Not all chemicals or scented products are bad. In many cases there might even be an improvement over products of past years. If you have concerns, it might not hurt to stop using certain products altogether. After all, if you don't like what's on TV you have a choice—you can turn it off.

PERFUMES AND FRAGRANCES

Several hundred ingredients may be used in a single perfume. Even a partial list of these chemicals sounds like findings from a toxic waste site. Chemicals with names such as cyclohexanol, galaxolide, methyl ethyl ketone, linalool, benzyl salicylate, vertenex, geraniol, coumarin, musk ketone, musk ambrette are the ingredients of a perfume. Nothing against a little perfume, but let's take a look at the last two in this short list: musk ketone and musk ambrette.

During a year when such data were recorded, 1987, musk aroma reportedly accounts for five percent of the volume of fragrances and ten percent of their monetary value. A total of seven-thousand tons of various musk aromatics were used in fragrances. The three most commonly used musks are the aromatic chemicals: musk ambrette, musk ketone, and musk xylene. According to the National Academy of Sciences (NAS), there is virtually no testing for neurotoxic effects of fragrance chemicals. And guess what, musk scents are neurotoxic.

What makes something neurotoxic? When it affects the nervous system. For example, research published on musk ambrette indicates that this chemical causes "central and peripheral nervous system damage." Others in the above list have been found to irritate the upper respiratory tract and depress the immune system.

Furthermore, according to the NAS, fragrances should be included in the same grouping along with heavy metals, solvents, insecticides, food additives, and certain air pollutants. This is some heavy company.

Cyclohexanol was found to have a narcotic effect somewhere between benzene and chloroform when inhaled.

Aroma chemicals are not the only perfume ingredients capable of causing difficulties. A number of "natural" materials are used in perfumes to which hypersensitive people can be expected to react adversely. These are essential oils and include civet, galbanum, patchouli oil, and asefetida as well as bergamot oil. This latter is classified as a hazardous substance and a strong sensitizer on the same order as formaldehyde. Its use in perfumes is increasing.

Approximately eighty percent of perfume industry revenue is derived from perfuming objects ranging from plastics, fabrics and clothing, rubber tires and automobile interiors, to tobacco, soap, and cleaning products and medicines. To this list we can add perfuming by mail. The FDA has declined to take action thus far since "the number of people experiencing adverse reactions to perfume is still very small..."

At one time virtually all department stores added fragrance to their envelopes and bills. It is still a continuing problem. Not only do we get to smell it, but it is in the glue that we lick to seal the envelope. If you will recall the postal service had flavored stamps for a number of years. Then they listened to complaints and developed the self adhesive stamp.

We don't just get to smell the fragrance, we get to lick it. Headaches and contact dermatitis are two of many symptoms that have been associated with fragrance strips. In a laboratory, asthmatic patients have been found to develop symptoms of chest tightness, shortness of breath, wheezing and rhinitis after smelling the strips. This is high powered marketing that is not related to better breathing. We don't seem to have any choice in the matter, or do we?

One department store sends you a bill with an advertisement for "....the only fragrances with a synthesized human pheromone component which promotes attractiveness by enhancing the way we feel about ourselves and how others respond to us." This is actually written on the envelope that is used to send in your payment.

Pheromones are sexual attractants. They are produced by humans, animals and insects. Pheromones are released by sweat glands and are thought to be detected in an area at the back of the nose. Now they are added to perfume.

Many national catalogue companies stopped the use of fragrance in mailings a few years ago after public pressure to do so.

Fragrances in general are unregulated and have one or more toxic components. In fact, the National Academy of Sciences has noted that there is virtually no testing for nervous system effects of fragrance chemicals. This same agency found that depression of the immune system can occur with the use of some of these fragrances. The agency stated that virtually all fragrances are untested and many of those that have been tested belong in the same grouping as solvents, insecticides and food additives.

Fragrance samples may be hazardous to your breathing and to your general health. Many magazines contain fragrance strips as a form of advertising. These strips are prepared by dipping a paper strip into a scented oil so that an odor is emitted. The reader has no choice but to smell it. Bills, inserts and return envelopes are commonly scented.

According to the National Center for Environmental health Strategies (NCEHS) big industry tried to set standards for the use of state-of-the-art technology with the fragrance encapsulated "as completely as possible" as discussed below.

There is real fragrance and there is synthetic fragrance. It is the synthetic fragrance that causes the most problems. There is also free oil and there is encapsulated oil., i.e., oil in micro-capsules. The free oil gets into the air more readily and thus is more fragrant. The industry position calls for "no free oil" to be added to such inserts. What industry wants to do is to encapsulate the oil. The is unacceptable. While this may be an improvement, proposals to encapsulate fragrance strips are not adequate to protect those who are allergic, asthmatic or chemically sensitive.

Several national magazines have apologized to readers after one-time scent strip advertising when numerous complaints were received Also, it's okay to say no to department store persons who want to spray you with their latest perfume product. The Food and Drug Administration (FDA) has received an increasing number of letters from people who have been made ill from fragrances over the last several years. The FDA acknowledged that "It appears that the incidence and severity of adverse reactions has been increasing. We have learned that the people adversely affected by certain perfume odors also react to other strong odors."

AROMA THERAPY

A new fashion, one that has been around for some thousands of years, we call an emerging science. For example, there is an emerging science of aroma therapy—the use of fragrances and scents to cure a wide variety of physical and mental ailments. The treatment has been claimed to cure stretch marks, is an antidepressant, a sedative or stimulant. You name it. But there is more to talk about in terms of potential respiratory problems.

Let's examine the pros and cons of a couple of examples. Pinenes are oils present in pine, nutmeg and black pepper. They belong to a class of chemicals called terpenes. Pine oil has been recommended as a treatment for sore throats, for "respiratory weakness," to treat various infections and to relieve stress.

Another terpene is limonene, present in bergamot and the main constituent in many citrus oils, including lime, lemon and orange. It has been recommended as a stimulant to the immune system, for treatment of herpes and as a stress reducer, among other things. Both pinenes and limonenes have antimicrobial properties. As you might expect, they smell like pine and citrus. Does this mean that they are safe to inhale?

Chemically sensitive (CS) persons appear to be a new breed, born in an age of chemical toxins. The methods of choice for aroma therapy are bath, compress, massage and also diffuser. (A diffuser is a device that spreads the scent of essential oils into the air without the inconvenience of burning them.) As natural as these fragrances and methods of application might be, persons with CS cannot tolerate them.

Inhalation therapy can have adverse effects on asthma patients. The herbs recommended for treatment of asthma include cajeput, cypress, eucalyptus, frankincense, hyssop, lavender, lime, myrrh, spearmint, spruce and others. These plant scents have been used for centuries and even thousands of years to lessen symptoms of asthma. This is at cross purpose with today's practice of avoidance.

Medical doctors tell us that many of these treatments will work to some extent with the right patient at the right time. Since some eighty-five percent of asthma is allergic, aroma therapy may be a problem for those patients with allergies to cypress, spruce and the rest. Is today's world so different from that of the past? Has western knowledge clouded our common sense?

Virtually all manner of paper products, soaps and personal care items are fragranced these days.

We are sprayed as we enter the department store. We are exposed to fragrance strips in magazines as well as ads in the mail. Many of the scents are toxic to the nervous systems and are unregulated in their sales and distribution. They can affect our attitudes and behavior.

Here's the point. Whether or not you believe you have some sensitivity to fragrances and perfumes it is less irritating to the respiratory tract to avoid contact with these substances as much as possible. Buy fragrance-free paper products (recycled products are usually fragrance-free) and are brown in color; read the label and buy fragrance-free soaps and detergents.

COMMON LUNG IRRITANTS

Let's look at the chemicals that are in some household products that have been around for awhile and that aren't so safe. First some definitions: Emulsifiers keep substances together in solution that might normally separate, such as water and oil.

Petroleum distillates (including mineral spirits) are colorless solvents obtained from the distillation of petroleum. They are toxic and flammable and serve as good grease cutters.

Surfactants are agents that lower the surface tension of liquids to allow for better penetration for cleaning purposes.

Air fresheners contain essential oils and aromatic chemicals. Many contain formaldehyde, perfume and dye.

Some can decompose into chlorine and phosgene. They only cover up an odor; they do not remove it. They are most annoying when used in public places such as health clubs and rest rooms or in advertising fragrance strips. If you are serious about fresh air then don't use air fresheners. More about this later.

Dishwasher detergents are powders made of strong alkalis and phosphates (or sodium carbonate). They also contain bleaches, perfumes, colorings and other ingredients. The entire mix not only clears away the grease from dishes but prevents them from spotting. Their greatest user-hazard comes from inhaling the powder dust when the soap is added into the dishwasher. Thus, liquid soaps might be preferred by some.

Dust control sprays are composed mostly of mineral oil, alcohol, petroleum distillate, surfactant (to dissolve grease), fragrance and propellant. If you like air fresheners you should love these sprays.

Fabric softening sheets are non woven sheets of rayon fibers saturated with surfactants and perfume. These work as moisturizers to prevent build-up of static electricity in the clothing. The perfumes can be irritating to the eyes, nose and throat and cause headaches and dizziness in persons exposed to them for short to long periods of time. They are also absorbed into the clothing and bedding. When used with fragranced laundry detergents a potent combination can be obtained.

Furniture polishes contain water, petroleum distillate, emulsifier, wax and possibly fragrance. The use of adequate ventilation is the key to avoid getting into trouble with these products.

Mothproofing products with paradichlorobenzene are less toxic than those with naphthalene, but overexposure to either can still cause dizziness. Try to air out the clothing the night before wearing it to minimize skin contact with these chemical irritants.

Varnishes commonly contain linseed oil or castor oil, drying agents and petroleum distillate. Similar to furniture polishes, these products are frequently used over large surface areas. There should be adequate fresh air ventilation to prevent irritation to the eyes and lungs. And don't forget to properly dispose of any rags that are used. If stored indoors these rags will serve as a fire hazard and as an annoying and irritating source of odor, long after the varnish or polish on the furniture dries.

AIR FRESHENERS

"Air fresheners" are big business. They come in solid, wick (liquid), spray and plug-in styles. They are available in a wide variety of scents and are used in public and private restrooms, gyms, restaurants, bars and cars. Usually air fresheners will contain essential oils and aromatic chemicals or water and propellant. Except for the water the other ingredients are in question. They can include naphthalene (banned as a moth repellent due to its carcinogenic properties), phenol, cresol, methylene chloride, para-dichlorobenzene, ethanol, xylene and formaldehyde (a carcinogen). At the minimum they will contain a perfume masking agent.

According to the National Institute for Occupational Safety and Health (NIOSH) and other sources phenol is an eye, nose and throat irritant and has a sweet and pungent odor. Cresol also has a sweet-type odor and affects the central nervous system.

Some air fresheners contain methylene chloride, a paint stripper and carcinogen. It's listed by NIOSH as having a chloroform-like odor, and eye and respiratory irritant at higher concentrations. Air fresheners also commonly contain para-dichlorobenzene (p-DCB), another carcinogen. It has a mothball-like odor.

This latter is most common in solid and aerosol air fresheners. It is used as a fumigant, insecticide and germicide. In nearly every home found to contain p-DCB, an air freshener was present. Both methylene chloride and p-DCB may cover up a smell or affect the nose such that you lose the ability to smell.

Air fresheners are highly advertised but serve no real function in our daily lives and very little research has gone into their effects on health. Information is lacking. Since ingredients don't have to be listed, we have no way of knowing which air fresheners are safer than others. Even the added fragrance is unregulated and untested, in keeping with the perfume and fragrance industry as a whole.

The use of essential oils in air fresheners is questionable. "Natural" essential oils come from plant sources and smell like the plant. Lemon is one example. Artificial oils or perfume oils are made from petrochemicals. These may act in combination with the other ingredients to have a negative effect on respiratory health.

Instead of covering up household odors, get rid of them at their source: Use one-half cup of borax in the bottom of the garbage can, grind a lemon peel in the garbage disposal, run the bathroom and kitchen fans more often, check the refrigerator for rotten food, moldy cheese, fermented jams and jellies.

Wash dirty socks, underwear and sweaty gym or yard clothes on a regular basis.

Keep your pets clean, and change the litter box frequently. Otherwise, cats will track their "litter business" all over the house, even onto kitchen counters. Airing out the home on a regular basis is one way to provide fresh air without trying to create it.

If you like a scent in the home, natural scents such as fresh flowers, pine or natural herbs such as spearmint are readily available in the marketplace. If you chose to purchase a fragrance product, first make sure you don't react to it before bringing it into the home.

The off-gassing of the chemicals from "air fresheners" occurs at very low concentrations. This does not make them safe. It just means that health effects are unknown at the lower concentrations. This is no excuse for their usage in a product that is sold in tens of millions of units each year.

It is necessary to understand that combination effects are not only possible, but likely. This means that the effect on the human body can be from ten to one thousand times greater that either chemical used alone.

The third thing is that there is little or no state or federal regulation regarding what can be put into a product and sold for home usage, Hence, nothing has be written on the label. If air fresheners were safe then manufacturers would list their ingredients.

Essential oils are concentrated oily substances extracted from plants. They are used in perfumes, antiseptics and antibiotics. No prescription is required. Most of us are familiar with the scent of eucalyptus or menthol to clear up congestion. The smell of essential oils is not recommended for asthmatics or persons who are sensitive to chemicals. Their use as one of many ingredients in air fresheners is questionable.

Essential oils have a complex chemistry and can have a large number of ingredients. Typically they contain terpenes, esters, alcohols, phenols, aldehydes, vitamins and hormones such as a estrogen (sage). This latter is not recommended for persons who are trying to avoid estrogen-like chemicals.

PESTICIDES AND OTHER HAZARDS

Most homes studied had up to thirteen different pesticides indoors. Many of them had never been used there.

Pesticides used in the home include Heptachlor, Chlorpyrifos, Aldrin, DDT, Chlordane, Diazinon, Atrazine, Dieldrin, Carbaryl and others. DDT was banned in 1972. A 1992 study in the midwest found that twenty-five percent of the homes tested had DDT in the carpets. This was twenty years after DDT was banned. And according to well documented reports DDT is present the blood of most middle age or older persons.

The level of indoor pesticides is at a level some ten times or higher indoors than outdoors in virtually every study conducted. This includes pesticides approved for outdoor usage only. Most home studies had an average of thirteen pesticides indoors which got there because of several factors: we sprayed them there; the outdoor types were exposed to warmer air, evaporated and entered the home as a gas to then bind with the dust; we tracked them inside on our shoes. Pesticides can live for years indoors where they bind with dust in protected areas away from sunlight.

Look at benzene. Nearly eighty-five percent of atmospheric benzene is produced by cars burning gasoline; the remaining fifteen percent is produced by industry, such as petroleum refineries. But studies have found that about three percent of our total exposure can be attributed to industry. The entire contribution to atmospheric benzene of six hundred billion cigarettes smoked annually in the United States is approximately one-tenth percent of the total benzene put out by cars and industry. Yet nearly all of our exposure to benzene is from indoor sources. This is because indoor smoking traps the chemical in the dust. And outdoor dust that contains benzene comes indoors.

Combustion, or burning, yields fine particles that are extremely dangerous for a number of reasons. Smoking, unvented cooking, burning candles, firewood or kerosene leads to the formation of enough benzo(a)pyrene indoors to equal the inhalation of three cigarettes per day by an infant.

Learn to steer clear of benzene, petroleum distillates, "inerts" and indoor pesticides, whether they are floral scented or unscented. The effects of these chemicals on health and welfare may be made worse by exposure to a high level of indoor ozone.

There are recommended methods to reduce the level of lead in children's blood, as well as exposure to indoor allergens, particles, and VOCs. Avoiding the use of mothballs and air freshners with p-dichlorobenzene reduces the risk of cancer. Avoiding storage of gasoline and lawn mowers in attached garages means less risk of fire and of cancer from benzene. The numbers are in.

Newly dry-cleaned clothes take about a week to lose these residues by either wearing them or by off-gasing in the home. In either case we inhale or absorb the chemical. It is a carcinogen. This problem can be cut down only by some twenty percent by airing the clothing for a day outside before moving it indoors.

We also have numbers to come later, regarding how much of this carcinogen is released in the car when you are hauling around the clothes.

The finding that there is lead in household dust is a major discovery. The amount can vary from low to high in the same or different homes. Infants and children are more likely to get lead poisoning than adults due to the continued development of their nervous system. Infants crawl on rugs and carpets with dust (and lead), then put their fingers and hands in their mouths, transferring the lead. It is quite likely that more investigations will find that adults also have problems caused by this element. An example here is increase in blood pressure associated with lead exposure.

It has been found that low-level exposure to lead causes a decrease in IQ. Lead exposure has also been found to tie in with speech and language problems, increased aggressiveness and acting out (in addition to VOCs), reduced attention and lower scores in reading, spelling and math. It has also been reported that there is no safe level of lead.

All this said, don't panic. It takes time to develop a logical cleanup procedure.

According to the EPA, some ninety percent of all households in the U.S. use pesticides. Some eighty-five percent of the total daily exposure of adults to airborne pesticides is from breathing air in the home.

Numerous studies have shown that pesticides can last for months, years, or as long as a quarter centuryindoors. This is because inside the home they are protected from sunlight, heat, rain and other environmental factors that normally would cause them to wash-our or degrade. Pesticides that have no known use at a home can still be found there because of tracking.

Carpet dust is the culprit for the site of capture for indoor pesticides. From absorption onto the dust particle they are released into the air or stay attached to the dust particle. Because of this there are several important considerations:

1. Pesticides get into the dust due to indoor spraying or tracking-in from the out-of-doors. The spraying is from commercial applications or private resident use.

2. Depending on the volatility of the pesticide, the amount inhaled in the air can exceed the amount breathed from the dust by four-fold to fifty-fold.

3. Infants and toddlers breathe in carpet dust and have more hand-mouth contact than adults. Also, dust containing pesticide can settle on other surfaces in the home such as counter tops.

Dieldren, a termiticide, was commonly used in the 1970s. Today it is frequently found in homes. This is also true of Chlordane and the major constituent of its formulation, Heptachlor. Pentachlorophenol is a pesticide that is so widespread that it was found in the urine of seventy-two percent of the population tested. Vapors from the active ingredient in Dursban, chlorpyrifos, as well as Chlordane and PCP enter the respiratory route at a much greater rate than ingestion of the dust itself.

If there is any good out of this, nobody knows how much in-home exposure is bad for us since long term studies are not in agreement.

What does all of this mean? Pesticide use in the lawn and garden and indoor treatment for bugs and termites result in the presence of these chemicals indoors. They can last indoors for months or decades. Handling plants that have been sprayed will also contribute to skin and respiratory absorption of pesticides.

Recent reports have found that soap and water are poor agents for removal of pesticides and their carrier solvent from the skin and other surfaces. Studies have shown that the use of simple isopropyl alcohol can successfully remove pesticides from the skin. And leaving your shoes by the front door appears to be an idea with a lot of merit.

There are more pesticides and a higher level of them indoors than outdoors. This is true even when there is no indoor spraying. Pesticides for outdoor use include insecticides, rodenticides, termiticides, herbicides and mildewcides. These pesticides enter the home on shoes as they are tracked in from the lawn or garden or neighbor's yard. They are present when we buy and use aerosols, fogs, sprays and powders and apply them in corners, edges, floors and in the air. They are present when commercial companies apply them to baseboards on a monthly or quarterly basis. They are present on vegetables, fruits, grains, meats, and some clothing. An example of this latter is when a cotton crop is sprayed to hold down insect infestation and the cotton is made into yarn for clothing.

Any kind of heat causes chemicals to become airborne. This includes sunlight, central heating or normal day and night temperature changes. The chemicals then settle onto colder surfaces such as the outside of appliances, mirrors and windows. This usually occurs at night. As these surfaces in turn warm up the chemicals become airborne again to settle into carpets, furniture, bedding, window coverings and other soft furnishings. They can also become attached to wallpaper.

Throughout this process a number of things happen: the chemicals are spread throughout the home by the ventilation system; they either stay in the gas state or become tied to dust particles. Also, children crawl in the dust or come in contact with the surfaces that have the pesticides.

In the carpet, the pesticides work their way into the backing and the pad underneath. Both of these act as long-term reservoirs that transfer the pesticide to dust. Pesticides then re-enter the air as we vacuum, and from heat-caused evaporation (volatilization). Older vacuum cleaners with poor quality bags are the worst in terms of permitting dust leakage.

The people most susceptible to pesticides are infants, children, pregnant women, the elderly, people taking medication, those persons under immuno-therapy or those with impaired organ system functions.

Studies are showing that repeated exposure to low doses of pesticides can cause chronic illness, just as can high doses over a short period of time. A report issued by the New York State Department of Law notes that "exposure may occur at the site of pesticide application or elsewhere as the result of airborne movement, tracking, or routine cleaning activities. Pesticides may persist for days, weeks, months, or even years after indoor application."

The United States General Services Administration has established a successful program for Integrated Pest Management for thirty million square feet of government offices. We should do the same for our homes. Basically, this means that if we are concerned about our health we should treat problems as they come up, and not use the shotgun approach.

There is a group of pesticides that we encounter quite often. They work very well against cockroaches, ants, flies, lice, mosquitoes, caterpillars, aphids and other insects. Also, they are much less toxic than most pesticides, unless you have allergies.

They are related to materials derived from the chrysanthemum flower and as such, belong to the ragweed family. The use of this flower against insects dates back to the early Persians and Chinese. Today the major sources are Tanzania and Kenya.

Pyrethrum refers to the dried, powdered flower heads of the plant. Pyrethrin is made from the actual ingredients in the flower that are active against the insect pest. Pyrethroid refers to man-made pyrethrins. These are much more toxic to insects than the other two.

These three insecticides are much less toxic than other pesticides on the market. They are available as a powder, liquid concentrate or aerosol. As a liquid they are in a water base. Aerosols are the most dangerous to humans.

The propellant used these days is carbon dioxide since fluorocarbons were taken off the market.

It is the application of a pesticide or herbicide in an aerosol that is the problem here. So much is released into the air that there is a great hazard to respiratory health. This is especially true when used indoors. This can get tricky when manufacturers include fragrance into the formulation. We inhale the sweetness of the perfume and never realize that we are inhaling poison. And don't be fooled into thinking that it is harmless to humans because it is used against insects.

Breathing these compounds will result in their immediate absorption into the bloodstream. This can cause immediate and severe effects to millions of ragweed sensitive persons. Using the liquid or powdered form is much less hazardous to your health or that of your family and neighbors.

The presence of this class of pesticides may explain an allergy mystery. An elevated level of antibodies to ragweed can be found in the bloodstream of many persons when ragweed is not in season. It is possible that these persons have been exposed to the pyrethrum class of pesticides during or prior to the time they were tested.

These insecticides act almost instantly against the nervous system of pests. This is in contrast to boric acid which can take days to act. In order to ensure that these pesticides retain their knock-down power other slower acting ingredients are added to the mix. If you are going to use these compounds for treatment of fleas in your cat be aware that they can be toxic to cats at small doses. Read the directions carefully.

The man-made synthetic pyrethroids are coming into wider use against termites. The advantage these materials have over chlorpyrifos (now commonly used) from a health point of view is that they degrade more rapidly in the presence of sunlight. So if some of it gets loose it won't last for months or years, as does chlorpyrifos.

Pesticides have long been used for bug and weed control at home, school, the work place, roadways, agriculture and community-wide applications. Pesticides in lawn care is big business.

Pesticides are comprised of two primary components. The first component is the active ingredient. This active ingredient kills or retards the growth of a particular group of bugs or weeds. The second component consists of other substances. Listed as "inerts," these miscellaneous poisons make up eighty to ninety-nine percent of the pesticide formulation. Inerts include solvents, stabilizers, emulsifiers and preservatives.

What you need to know is that when a substance such as DDT is banned as an active ingredient because of its tremendous danger to health it can be used in much higher amounts as an inert, and as such, does not have to be listed as an ingredient on the label.

Other inert ingredients in the over 50,000 pesticide products on the market today include naphthalene, xylene, toluene, glycol ether, trichloroethane and dioxin. Recognize any of these names? Oh, inerts are virtually unregulated by the EPA. Just another reason not to have a lawn.

Here's another reason. Dogs exposed to the herbicide 2,4-D are up to twice as likely to develop malignant lymphoma according to a study by the National Cancer Institute. 2,4-D is in over 1500 pesticide formulations. Lymphoma among Americans has increased by about fifty percent since 1973—one of the largest increases of any cancer.

To make matters more interesting, when a pesticide is used in some other function, say as a room deodorizer, the chemical does not have to be listed on the label.

Paradichlorobenzene is one example and is used in air fresheners. This pesticide is not only highly toxic to persons described as chemically sensitive, but seriously affects "normal" persons to the extent that it many contribute to their eventual sensitivity to chemicals.

Traces of herbicides have been found in rainwater in twenty-three states, mostly in the Midwest and Northeast. These included alachlor and metalachlor. This is the first study to indicate that herbicides applied to farmlands can vaporize into the atmosphere and travel airborne for several hundred miles. This information also tells us that concentrations will be much higher in areas next to its application, such as inside of your home.

Finally, in a recent interview of 6800 persons claiming to be chemically sensitive, conducted by a patient group called the National Foundation for the Chemically Hypersensitive; nearly half of the patients say their illness started with a pesticide exposure and a large percentage of that group consists of teachers.

The topper is that name brand pesticides that are used outdoors and indoors against ants, cockroaches and flying insects as well as other pests. You hold the can some twelve inches away from the insect and spray. But you need to avoid breathing the pesticide that can also cause eye and skin irritation. They come in numerous scents.

The main question is this: What happens to the airborne pesticide when it circulates throughout the home? In a few words, the pesticide enters the air of the home. It is absorbed into the home's soft furnishings such as bedding, carpets, arm chairs, sofas and drapes where it will off-gas over time; it attaches to dust particles, especially in the carpeting, where small children will engage in hand to mouth exercises and ingest it; the pesticide will be picked up by the airstream and circulated though the ducts into all the rooms. Then it will be inhaled by all family members. Pick your flavor.

FORMALDEHYDE

Formaldehyde is a simple molecule that is frequently combined with urea, another simple molecule. The complex is a modern marvel with numerous properties that fit nicely into modern living and is found in such a wide variety of everyday products that the list boggles the imagination. It is a common respiratory irritant and is carcinogenic in extremely high concentrations. Under ordinary exposure conditions, however, the effects of formaldehyde on the human body is arguable.

Formaldehyde is one of the most misunderstood chemicals of our time. It is used in foam insulation, plywood, chipboard, textiles, paper, plastics, resins, paints, glues and cosmetics. It is used in the making of shoes, carpets, clothing and automobiles. It is found in dentistry, dyes, adhesives, leather products and paper.

Formaldehyde is a very soluble chemical. This means that it dissolves in the moist mucous of the nose and throat and rarely penetrates the mucous to irritate the nerves underneath. Its odor is pungent in small amounts but it is dissolved in the body within a minute and a half and turned into carbon dioxide and water. Basically, formaldehyde stays in the upper respiratory tract. The exception is smokers who inhale it into the lungs.

We usually start to smell formaldehyde at about 0.1-3.0 parts per million (ppm), more or less, depending on the person. Eye irritation follows smell. It is found in most homes at 0.03 ppm, in mobile homes at zero to 2.0 ppm, and in offices and day care centers at 0.4-0.6 ppm. Smokers inhale formaldehyde at 5-8 ppm, but sidestream smoke contains it at the level of 30-40 ppm. It is also emitted from wood burning fireplaces at low to high concentrations indoors. At the 8 ppm level it will penetrate the mucous lining of the respiratory tract and cause irritation and distress.

Formaldehyde is present in clothing stores at 0.9-3.3 ppm. Smoke from a log burning fireplace will contain a lot more formaldehyde than smoke from non-wood fires.

It is common to blame the smell of new carpets on formaldehyde. In most cases the smell is due to other chemicals present in the adhesives that hold the carpet surface to the backing or the carpet backing to the floor.

There are some unknowns about formaldehyde. For one thing, very little is known about its combination health effects when in the presence of other chemicals. Does one plus one equal two or ten? For another, people who state that they are chemically sensitive will exhibit a variety of nerve and muscle related symptoms when in the presence of extremely small amounts of the substance. These symptoms include headache, nausea, dizziness, fatigue, disorientation and loss of short term memory. The symptoms can last from minutes to days or longer.

It is known that formaldehyde is a 'sensitizer', similar to glutaraldehyde and Bergamot. In medical terms this means that exposure to enough formaldehyde can cause the body to become sensitized to other chemicals. This is similar to what occurs when overexposure to one pollen or mold type sensitizes the body to many pollen and mold types. The difference is that in one case the body becomes sensitized to chemicals and in the other the body is sensitized to allergens.

If you are concerned about formaldehyde in your home or business for whatever reason remember that most of the chemicals used indoors out-gas from their materials quickly over a period of days to weeks with lots of fresh air. After this they usually out-gas slowly in low amounts for months or years.

Exposure to formaldehyde is virtually inescapable. The energy crisis of the 1970s led to measures for tightening up buildings and reducing the rate of exchange between indoor and outdoor air. This practice, coupled with "improved" product technology led to an increase in the indoor air concentration of formaldehyde.

Products containing formaldehyde are carpets, wallpaper and textiles. Modern furniture is frequently constructed of particle board interior of wood veneer on the outer surfaces. Wallpapers can emit formaldehyde, especially prepasted types and those consisting of fibers or layers of paper bonded with formaldehyde containing resins. Add to the list creaseproof and flame-resistant and shrinkproof fabrics. Add curtains of cellulose acetate and bedsheets. Want more? Starch based glues, room deodorizers, shampoos, and cosmetics can contain urea-formaldehyde (UF) resins. How about paper products (where it is used to add wet strength). Paper products with UF resin include grocery bags, waxed paper, facial tissues, napkins, paper towels, and disposable sanitary products. UF resins are used in fabrics to provide shrink resistance, permanent press, water repellency fire retardance, and fixation of color. Compared with other formaldehyde emitters in the home, however, the amount of this gas emitted from paper products is probably quite small.

Formaldehyde toxicity is brought about by contact with skin and the mucous membranes of the eyes, nose, and throat. At extremely low concentrations formaldehyde can produce coughing, constriction in the chest, and wheezing. The level of formaldehyde in room air depends on the rate of exchange of air, the area of the emitting surface, the total air volume, and other factors such as humidity and temperature and the age of the source. The concentration of formaldehyde in the air is measured in parts per million (ppm). The amount that can react with a person depends on the sensitivity of the person and the length of exposure time.

It is important to remember that formaldehyde is readily absorbed by moisture; whether it be in the air, in the lungs, nose, eyes, or skin. Heavily debated is the finding that repeated exposure to formaldehyde liquid or vapor can cause certain individuals to become sensitized. Upon re-exposure to formaldehyde or to related substances (and sometimes to apparently non related substances) these persons may exhibit allergic dermatitis or mild to severe asthmatic reactions.

Urea Formaldehyde Foam Insulation (UFFI) homes were found to have from three to twenty times the formaldehyde level as non UFFI homes and reached levels of 0.2 ppm and above. However, it is mobile homes that are repeatedly the worst offenders in terms of releasing formaldehyde gas. Here the air concentrations range from 0.1 to 1.0 ppm with the kitchen areas the worst followed by the bedroom. Four-fifths of persons studied in one large mobile home survey were exposed to enough formaldehyde to irritate the eyes and upper airways. Memory lapse or drowsiness were common symptoms as was acute depression. Particle board and plywood are the major culprits here since counters, cabinets and paneling and other woodwork is made from it.

The people most easily and commonly affected by formaldehyde gas are those who already have respiratory disorders, the elderly, the infirm, and children. When you go shopping you should ask questions and read labels. Why? According to the National Institute for Occupational Safety and Health formaldehyde can cause: irritated eyes and tearing, burning of the nose and throat, cough, bronchial spasms, pulmonary irritation, dermatitis, nausea, loss of consciousness and vomiting and possibly cancer.

Coating or painting a formaldehyde-emitting object is an easy way of retarding the emission. A number of commercially available vapor barrier paints are available for this purpose (ask your local paint dealer or see the Appendix. As a general rule, it is a good idea to use plenty of ventilation while painting. Use two coats.

Emission can also be reduced by using mylar or vinyl wallpaper with a heavy canvas backing. If you suspect medical problems are occurring because of some furniture product that may be emitting formaldehyde, use plenty of open window fresh air to dilute the gas until you decide what to do.

Ideally, getting rid of the source is the best way to go. Since the out-gassing rate of formaldehyde is dependent upon temperature and humidity, simply lowering the thermostat and not using a humidifier can significantly reduce emissions.

Concern over protection from formaldehyde gas also applies to attic or crawlspace insulation that may have been installed within the past few years. That is why it is a good idea to find out exactly what was installed and to have any joints in the attic duct work inspected for breaks. Otherwise, formaldehyde gas may be constantly entering the home proper through your heating and cooling vents. Examine your symptoms and if necessary, consult with your doctor.

Of the formaldehyde based glues, there are two types: urea-formaldehyde and phenol formaldehyde. Of the two, the out-gassing characteristics of the former are 10-20 times greater than the other. That means it is much more toxic.

If you are constructing your own chemical free home, avoid plywood, particle board and carpet as a general rule.

According to the Weyerhaeuser Company and other sources, reduction of much of the formaldehyde in dwellings and especially in mobile homes can be accomplished by using ammonia fumigation. The ammonia odor disappears completely within several days. Ammonium hydroxide (standard ammonia solution) reacts with formaldehyde to create a stable compound known as hexamethylene tetramine. According to some literature sources this may be accomplished by simply spraying the home with ammonia solution. According to other sources more permanent relief can be obtained by using a concentrated ammonia solution that is left in a warm home for at least 12 hours.

For personal care items such as deodorants and hair care products read the label. For wood products avoid particle board, chip board and plywood. Ask questions. Wash new bed sheets before using them. Some air cleaners claim to have the ability to remove formaldehyde from the air. If they contain activated charcoal they may or may not be effective since only specially treated activated charcoal removes formaldehyde from the air. Try the unit before buying it.

Remember that formaldehyde affects the upper respiratory tract when breathed and commonly affects those persons with allergies. Be aware of the symptoms of breathing formaldehyde and the most common products in which it is found. (See book reference no. 9.)

The odor and toxic effects of formaldehyde, varnishes and other volatile structural and home repair materials is well known. These products take weeks or years to outgas (off-gas). Along with carpets, they are largely responsible for sick buildings, sick schools and sick homes.

Kitchen and bathroom cabinets and counters are usually made of pressed wood or particle board in homes, apartments and buildings. In schools they comprise the cabinet structure and affect teachers working in closed spaces.

Mobile homes are noted for their high concentration of formaldehyde which can be as high as one tenth part per million throughout the interior of the structure and a full one part per million in the kitchen area. In the kitchen a high percentage of structural material is made of textile products and artificial wood. Panelling is also present in mobile homes to a high degree and is of a synthetic wood-resin mixture coated with an out-gassing varnish.

Source control, ventilation and filtration are three methods to deal with this problem. In the first case, source control can include a method to seal in the chemicals and prevent their escape into the breathing spaces.

One can now obtain a non toxic sealer for your kitchen and bathroom cabinets, dressers, paneling, as well as other similar household items. The product is called AFM Hardseal and is produced by a company based in California (see Appendix). One quart should be enough for the average kitchen. When you use this product be sure to paint the inside of your cabinets since the inner cabinet area can emit toxic fumes for years. This is because it is closed to the fresh air and can be absorbed into your plastics, styrofoam products and dish towels.

The second method for dealing with an out-gassing problem is ventilation. Plenty of outdoor air, not recirculated air conditioned air, is the true key to ridding your airspace of unwanted fumes. It won't happen overnight, but it will happen.

The filtration method of gas removal involves a number of methods which include activated carbon and some very new, inexpensive, high tech filters. Filtration is discussed elsewhere in this book.

SEWER GAS

The smell of sewer gas has been experienced by virtually everyone. How does in get into our buildings and is it toxic?

First of all, the smell of sewer gas means that there is decomposition of organic debris by bacteria. A number of gases are released from their action and one of them, hydrogen sulfide, smells like rotten eggs. Hydrogen sulfide is "toxic" depending on your definition of the word. The fire department may tell you that it is not toxic. This means that it won't have any lasting effects after you get lots of fresh air. But our definition of the word may be a lot different. Burning eyes, coughing, nausea, headaches, disorientation and sleeplessness means that the gas is toxic. And in some people, the effects can last from minutes to months.

To most folks, the nose is our early warning system and we can detect the smell of sewer gas at a concentration of less than 1/400 the official toxicity level. The problem here is that hydrogen sulfide fatigues the sense of smell quickly, as does ozone and certain fragrance products. Thus we are unable to detect the presence of the rotten egg smell or even other odors before long. Find some fresh air for a few minutes to free the sense of smell once again and return to the problem area. This should give you a better indication of whether the gas is present or has gone.

All toilets are vented and so is our sewage system. Manhole covers have holes in them and toilets are vented with pipes that connect the sewage system to the roof. This is to prevent the build-up of flammable and explosive gases. That being said, sewer gas can enter our buildings in a number of ways. In a commercial setting, the roof vent stacks can be located within feet of an air conditioner that also takes in fresh air. When the wind blows the wrong way the sewer gas will blow into the fresh air system of the air conditioner and be conducted throughout the building or only to certain areas of the building.

This does not happen in homes because the air conditioner recycles all the air with no fresh air make-up.

Here is something that commonly occurs in commercial buildings and schools. These buildings contain floor drains, built to receive floor water after mopping. These drains lead to the sewage system and have a curved piece of pipe which is known as a "P" or "S" trap; a bent pipe that is made to contain a certain amount of water to keep the sewer smell out of the building. After a period of disuse these traps will dry up and sewer gas will then rise into the building. Filling them with water again quickly solves the problem.

In the home there are numerous ways in which the gas can enter. One way is for the wind to blow toward your home when there is a manhole cover directly in front or behind it. Another way is most mysterious. When using the fireplace, whether it is gas or wood, the odor of sewer gas may enter the home.

When using an appliance that causes air and oxygen to leave the home, air and oxygen must be sucked into the home to replace it. Appliances include the fireplace, the clothes dryer and kitchen and bathroom exhaust fans. These latter two have hoods on them on the roof to keep out the rain and snow.

Now back to our first example—the fireplace. Remember those sewer vent stacks and the outlets for the exhaust fans on the roof? Well, the oxygen that has to enter the home to replace what is being pushed out will do so in the easiest manner possible. It can suck air through these roof vents and if the breezes are blowing just right, the air will come from the side of the sewer vent. Air can also be sucked through a window with the same result.

As far as the sinks, the "P" trap may contain what is known as a "plug" of debris. This is okay. But if you fill the sink and then drain it the plug will be forced down into the sewer system creating a vacuum in the "P" trap area. Now there is a clear line between the sewer and your sink. This problem can be solved by running the water for a minute or so after draining the sink to refill the trap.

In an apartment building an upper unit can fill and drain the sink which can cause the water in the lower units to empty out of their "P" traps and lead to a rotten egg smell in everyone's residence. A dry "P" trap can also occur in a sink that has not been used for a long time or newly repaired, or a floor drain this is present in the heater or air conditioner closet or elsewhere. Dry traps can occur in school floor drains when water has not been run into them for a long period of time.

MIXING BLEACH WITH OTHER PRODUCTS

The misuse of household cleaning agents has been reported in a variety of medical journals. One topic that is frequently reported is the household mixing of chlorine bleach with other products.

The chlorine and chloramine gases that are produced can come in contact with moist surfaces such as the eyes, nose, and especially the upper respiratory tract, as well as the skin. Hydrochloric acid and other acids are then formed and the tissues become irritated. Symptoms include cough, hoarseness, sore throat, headache, and tearing of the eyes. Difficulty in breathing and shortness of breath for months afterward can occur if too much is inhaled.

Household bleach contains 5.25 percent sodium hypochlorite. Household ammonia is five to ten percent ammonia in water. Separately, these agents have low potential for toxic inhalation injury. However, the combination of the two agents is much worse than the agents breathed singly. These chemicals may play a role in terms of making us more reactive to all chemicals in the future, similar to the effect of formaldehyde.

Basically, what happens is that folks will mix the ammonia with the bleach and frequently, add in some laundry detergent for good measure. Then they go about cleaning things in a closed room for a long period of time. Poor ventilation. Sound familiar? They ignore the early warning signs of eye and throat irritation or headache.

Believe it or not, according to published reports, some people think that their early symptoms are a sign of the "power" of the cleaning mixture they created, therefore it would be even better if they continue cleaning for a few more hours. Without proper ventilation people get gassed enough to warrant a trip to the doctor or the hospital.

Fortunately, severe exposures rarely develop into pneumonitis and related disorders. As you might guess the amount of irritation depends on the amount of gas produced, the exposure time and the water content of the exposed tissues (Are you thirsty? Are you sweating?). It also depends on the age and relative health of the person at the time of exposure.

One study reported 216 cases of exposure to chlorine and chloramine gas. Some seventy-six percent of the patients overcame the symptoms within one to six hours, with fresh air and cool liquids as their main therapy. A number of other patients had to be treated with oxygen and bronchodilators. The experience can be frightening; as shown in the following:

Clean air should not be artificially scented. Artificial scents usually result from commercial products that are toxic. John and Jane Johnson can tell us about it as they go through their day. Incidentally, the adventures of John and Jane can be seen in Anytown, USA, on any day at any time.

As Jane sprays her hair this Saturday morning she breathes in a healthy dose of aerosol propellant. Borderline asthmatic, Jane has to take time out to catch her breath. John, on the other hand knows better and stays away from aerosols. He uses a pump sprayer to add a little musk scent to his body. What John doesn't know is that musk is a toxic product that affects the nervous system. And it is unregulated, like most perfumes. This means his behavior and attitudes will also deteriorate and the smell will certainly not help Jane's asthma. If John had known this he would have suspected that it didn't apply to them, anyway, only to everyone else.

John and Jane go shopping. In the clothing store they are greeted with formaldehyde at ten to one-hundred times the level that people can detect. Formaldehyde is used as a fire retardant in clothing and for permanent press as well as other uses. Their eyes water, throats burn and they cough. They both developed a "space cadet" feeling. It must have been something they ate.

In the carpet store they are greeted by 4PC, the adhesive that bonds the top of the carpet to the backing. Strong stuff. One might wonder how the employees can take it.

The supermarket has such a great mixture of smells: fried chicken, fragranced laundry detergent, bleach and cleaning products. Their appetite is whetted. They buy a little of each and go to the department store. A handy-hostess gives them each a shot of perfume, whether they like it or not. They like it.

Then to the video store for a couple of the latest scents in plastic—soft plastic out-gasses more than hard plastic. They are lucky since the video store carries both.

John has to make a quick stop at the office. It's closed on Saturday. Three thousand square feet of new carpeting await him. The air conditioning is on to distribute the unfiltered odor throughout the entire building. He sneaks a quick cigarette.

Time to go home. John is a clean-freak. He begins to work on polishing the furniture with commercial polish. The level of petroleum base volatile organic compounds goes up ten-fold within minutes in the home. John turns on a central fan, thinking that will help. Jane is starting to get a headache.

"Think I'll clean up in the kitchen," says Jane after dinner. She tries a new mixture of ammonia and chlorine bleach, liberating chloramine gas and John calls the paramedics. "Not too smart," thinks John as they cart her away.

When last we left Jane and John, Jane had mixed ammonia and chlorine bleach to create a toxic gas. She has been admitted to the hospital where she is tended to by a bevy of doctors and nurses. They all sport latex gloves with cornstarch as a lubricant. The microscopic starch particles are also on their clothing as a fine dust. These particles carry the latex antigen and make one or two of the attendants cough. One of them sneezes. Jane's latex allergy becomes activated. The particles are carried in the air throughout the hospital.

Jane is wheeled to X-ray. Here we find the release of X-ray development chemicals, including glutaraldehyde, which sensitizes radiologists to chemicals in general. It is also released into the air near a waiting room to affect waiting patients. It is certainly accidental but, in fact, couldn't have been planned any better.

At last Jane is moved to a room. John follows her, thankful for the clean air. He does not know two things. First, the pleated filters in the air handling unit just collapsed from overload. They have been changed every three months for years based on manufacturer's recommendations. Only nine weeks have passed. Also, the hundreds of alcohol wipes used daily raise the indoor level of volatile organic compounds (VOCs) to 700 parts per million in some areas including emergency, pediatrics, microbiology and hematology. Open trash containers are common and closed, foot-operated containers cannot be found.

Construction is going on at this hospital. Several of the wings are being remodeled with plastic sheeting holding in the dust. The plastic sheets are not carefully placed and Aspergillus mold spores associated with the dust are released into the air and affect immuno-compromised patients in a wing down the hall. Many of them develop a fungus disease of the lungs and ears. Several of them later die.

Jane is released the next day. Accompanied by her husband they find their car that is near the loading dock of the hospital—beneath the first floor where all the diesel trucks set idling near a sign that says, "No Smoking!" The diesel fumes add a fragrance that enters the air handling system also located in the basement. Some fumes waft to the roof of the hospital. Here the fresh air intakes are located next to the exhaust vents so that the two become one.

John and Jane get home at last where John sneaks a cigarette. "Blow a little my way," says Jane, and laughs. John doesn't think it's funny. This is his only vice. He decides to mow the lawn. Since it's such a nice day, he opens the home to air it out while he operates the power mower. Grass particles and fumes enter the home. John finishes work and comes indoors, taking off his work clothes in the laundry area. He shakes them out before putting them in the washer.

In many ways John and Jane are quite average in that they do not have enough knowledge to make informed decisions. With a little more education about how their environment and their decisions affect their lives, then their lives could be made a lot easier..

CANCER AGENTS IN OUR AIR AND FOOD

There are a number of risk factors associated with cancer. These include early onset of menstruation, late entry into menopause and never having had a child or breast-fed a child. In all cases the estrogen level becomes elevated. Also, diets high in animal fat or alcohol also seem to increase risk. A small percentage (five percent) of women carry a gene that makes them more susceptible to the disease.

There are estrogen-like compounds in our environment, some natural and some synthetic. The natural ones are found in foods such as broccoli, cauliflower and soy products. These are short-lived in the body and actually help to reduce the estrogen level. The synthetic ones are found in chlordane, DDT, Atrazine, Methoxychlor, Kepone, plastics and aromatic hydrocarbons. The latter are found as components of petroleum and are inhaled readily. This includes gasoline, car exhaust and possibly petroleum distillates used in polishes and cleaners.

DDT persists in the environment for more than fifty years. DDT was banned in the U.S. in 1972 as an active ingredient but was used as an "inert" ingredient until the late 1980s. It is still used outside the U.S. as a pesticide and can be found in many imported crops. Methoxychlor is an insecticide used on trees and vegetables. Kepone was used in ant and roach traps until 1977.

We have all had direct contact with these agents. They are eaten by cattle in their diet of grasses and grains, and stored in the fat. The cattle are eaten by humans and the insecticides and pesticides are absorbed and stored in human fatty tissue. The pesticides are poorly degraded in the body and persists for years and decades. In the laboratory it was found that small quantities of two chemicals added together caused more cells to mutate than either chemical used at twice the concentration. This is the "one plus one equals ten" effect.

The fact that many of these agents are no longer used is irrelevant. Exposure builds up over time. New estrogen-like agents are being produced all the time. For example, similar estrogen-like compounds are released into the air when we cause plastics or food cans to get hot.

What to do? Actively promote research in the field of safer pesticides. On a daily level: 1) Avoid eating animal fat, since it stores pesticides 2) Wash fruits and vegetables thoroughly, since they are sprayed with pesticides 3) Be extremely cautious of the use of pesticides in and around the home, since they are inhaled and absorbed through the skin. This is especially true with infants.

DENTAL FILLINGS

Based on 1992 dental manufacturer specifications, dental amalgam (a mixture) contains approximately fifty percent mercury, thirty-five percent silver, nine percent tin, six percent copper and a trace of zinc. This formulation has changed somewhat over the 160 years that it has been in use. More than one-hundred million mercury fillings are placed each year in the U.S. alone.

The controversy to this story is that the mercury is implicated as a causative agent of Alzheimer's Disease since it localizes in the memory portion of the brain, damages this portion, and is found in higher amounts in persons with the disease. Its amount in the brain and other organs is directly related to the number of dental fillings. Mercury also stimulates autoimmune diseases such as arthritis in experimental animals, and affects the immune and nervous systems. In the latter case, mercury has been shown to cause decreases in attention and muscle function tasks among dentists themselves. The silver has also been found to be toxic in many of the same ways as the mercury.

It took all these years and a lot of high technology to even begin to gather some evidence. In fact, only in the past decade have researchers found out that mercury is continuously released as a vapor in the mouth, then goes to the lungs, is absorbed into bodily tissues, and binds to proteins. The vapor is in highest amount after chewing, and in people who have the most fillings. Mercury vapor is continuously released into the body.

Much of the mercury is swallowed with the saliva, and passes out of the body via the intestine.The rest passes out of the blood into the kidney where some is stored to impair renal function or to pass out into the urine.

According to the article cited above, there is no evidence to show that amalgam is safe. On the other hand we still need proof that it is not. Should you run out and replace all your fillings? That's your choice.

In terms of the number of exposed individuals, the most prevalent source of deliberate mercury exposure is almost certainly dental amalgam. It is difficult to link relatively small exposures of mercury from dental amalgam to specific diseases at this early stage in the investigation. However, research findings to date have shown that mercury vapor can penetrate the blood-brain barrier and the placenta. Since mercury vapor can pass into all cells of the body, it is potentially capable of affecting the immune system, sperm cells as well as the brains of unborn children.

Medical researchers have recently identified mercury as a possible cause of Alzheimer's disease and have stated the most probable source of the elevated amounts of mercury in the brain of Alzheimer's victims is dental mercury.

Various dental associations maintain that the mercury in fillings is localized and cannot escape into the body. Various citizens groups want their "silver" fillings replaced with plastic or porcelain. Some even maintain that this service should be available to everyone who wants it and let insurance pay the bill. The dental associations state that this is not cost effective, especially given the fact that there is no hard evidence for toxicity due to dental fillings.

Various states are starting to pass "right to know" and "informed consent" laws whereby the patient has to be told of the potential hazards of dental amalgam. This is now the case in California.

While the dental community itself is quite divided over this issue of toxicity of mercury and silver in dental fillings it would seem that we have two courses of action. First, the time is ripe for dentistry to push polymeric and ceramic materials for fillings. This is a less affordable option than silver fillings, unfortunately, because it was silver fillings that brought dentistry into the realm of the average pocketbook in the first place. The second option is to floss and brush to prevent cavities in the first place.

In the practice of dentistry, some ninety five percent of gold inlay work has silver-mercury beneath it, that is, in the cavity space. According to the texts (See Appendix) and dentists sensitive to this issue, this means that even if you have gold inlay, mercury is being directly absorbed into the nerve chamber, which leads to direct absorption of mercury into the blood stream. Chewing food or gum releases mercury from the filling into the oral cavity, and the vapor is then inhaled—an additional problem. For most patients, a silver filling lasts about eight years. Filling replacement involves the grinding of fillings with a drill that operates at close to one-half million revolutions per second. This scatters the silver (mercury) throughout the mouth and vaporizes the metal. The dentist can see the mouth turn dark in color because of the tiny silver-mercury particles. Absorption and inhalation of toxic mercury takes place at a high rate at this time.

Now that the old filling is drilled out the cavity is refilled. The replacement filling is packed much like cement. When it is packed, the mercury rises to the top, much like tamping wet cement which leaves the water at the top. Whether you spit out the mercury or the mercury is removed by suction there is an incredible amount of toxic mercury in your mouth that is absorbed into your system.

The EPA has limits regarding the rate that a person should inhale mercury from a toxic spill. As measured by probes designed for this purpose the amount of mercury vapor in some mouths can exceed this amount by as much as 25,000 fold.

INFORMATION IS AVAILABLE

There are complex federal requirements that regard the supplying of a Material Safety Data Sheet (MSDS) for toxic chemicals. But simply put, employers shall provide an MSDS to employees when a hazardous chemical is used in the work place. Usually, businesses will also provide an MSDS to a customer upon request. Paint stores and pest control companies are two common businesses that maintain one or more MSDSs on their products.

Now, the MSDS has a lot of scientific and medical jargon and it applies to the chemicals in a product in their purest form as well as in their final diluted form.

Take a look at fragrances and the MSDS. Fragrances frequently are used to mask other smells, and are supposed to have a "pleasant" smell of their own. One common type of fragrance is used in athletic clubs and is billed as an "odor neutralizer." Its purpose is obvious. The MSDS on this product states that it contains methylene chloride, ethyl alcohol, methyl alcohol and anisobutane and isopropane blend. The MSDS tells us that this product is non flammable, but the aerosols should not be exposed to temperatures above 130 degrees or the container may explode. So far so good. According to the MSDS, this particular product can decompose into hydrochloric acid (muriatic acid for swimming pool buffs), carbon monoxide, chlorine and phosgene. Phosgene gas is created when chlorine is mixed with carbon monoxide. It is a highly poisonous gas and is a severe respiratory irritant. The MSDS goes on to tell us that this product is a "simple" asphyxiant if vapors are trapped and inhaled, and over exposure to the "odor neutralizer" leads to headaches, dizziness, nausea, possible unconsciousness and even death. Or it can be a mild irritant to the eyes.

A second MSDS involves popular furniture polish, which contains our good friend, petroleum distillates as well as mineral oil. The MSDS states that it should not be mixed with other chemicals. It is flammable. It is not expected to be an eye irritant but may cause skin irritation.

Prolonged exposure to high vapor concentrations of this product may cause signs and symptoms of headache or dizziness. Affected persons "normally" experience complete recovery when removed from the exposure area. This product is not expected to be toxic by ingestion. However, if this material is swallowed and aspirated into the lungs, chemical pneumonitis may result. Excessive inhalation can cause headaches. Move the affected person to an uncontaminated area and get medical attention.

Finally, check the MSDS regarding paints you are looking at and buy the safest ones.

CHAPTER 5
MANAGING INDOOR AIR QUALITY

It is no secret that the rate of cancer and various respiratory diseases has been increasing in the United States.

Scientific American magazine (February 1998) discusses just this issue in an article entitled, "Everyday exposure to toxic pollutants." The article points to some sobering facts.

While we spend ninety percent of our time indoors and our exposure to toxins is five to ten times what it would be outside, our state and federal governments spend virtually all their air quality money for outdoor air pollution. The level of indoor pesticides is typically ten times higher indoors than outdoors in virtually every study conducted—including pesticides approved for outdoor usage only.

The average concentration of benzene, present in gasoline and household products (including some paints) and cigarettes, is three times higher indoors than outdoors.

The authors of the article ask this question: Could everyday items be more a health threat than industrial pollution, even when homes are surrounded by factories?Yes, is the answer.

The dry cleaning agent, tetrachloroethylene, is a carcinogen. Moth-repellent cakes or crystals, toilet disinfectants, and deodorizers are the major source of exposure to paradichlorobenzene. Both of these chemicals are carcinogens.

Learn to steer clear of benzene, petroleum distillates, "inerts" and indoor pesticides, whether they are floral scented or unscented.

Combustion yields fine particles that are extremely dangerous for a number of reasons. Smoking, cooking, burning candles, firewood or kerosene leads to the formation of enough benzo(a)pyrene indoors to equal the inhalation of three cigarettes per day by an infant.

We can divide indoor air problems into two main parts: gases and particles.

Gases include volatile organic compounds or chemicals. Another name for them is VOCs. These are chemicals that evaporate readily. This enables us to smell them, in most cases. They include acetone in nail polish and polish remover, the odor of mold, formaldehyde, furniture and floor polish, the thousands of products that contain petroleum distillates, hair spray and spray insecticides, hundreds of products that contain fragrance, new carpets and their glues, paints, varnishes and thousands of products that fit into these categories. They also include such harmless smells as brewed coffee and microwave popcorn.

Virtually all are lung irritants and will probably make our allergies worse if their level in the home is too high.

Almost every indoor environment will have some background VOCs. These can be measured at a level of 0.1-0.5 parts per million. This background level includes coffee; fragrance on clothing that has been washed and dried in fragrance detergent, fabric softener and anti-static products; deodorants and hair spray, natural human odor and a dozen other odors that we normally encounter in our daily lives.

It is mainly the petroleum based products that lead to trouble. These include petroleum distillates, pesticides, "inert ingredients", solvents and cleaners.

Clinical symptoms of over-exposure to VOCs include headache, itching and watering eyes, sore throat, coughing and even loss of short term memory. A host of problems can result from simple exposures to effects that can occur years later. In the end, chemical sensitivity may result.

Okay, enough with facts. Let's talk about how to get rid of excess VOCs.

The main problem with VOCs indoors is that they build up in concentration because we don't open the home or office to air it out on a regular basis. It's as simple as that.

The indoor VOC level can reach as high as 10-60 parts per million. (The typical home garage registers about 5-10 parts per million.) This can result from an airtight home or business that does not have enough fresh air. The good news is in almost all instances, the level of indoor VOCs can be reduced.

There are several remedies for a high or suspected high VOC level. First, air out the home or business structure regularly. Open the windows and doors. Even twice weekly in the morning with table top fans or central air fans moving the air will help drop the level of VOCs and complaints.

Second, wash the walls, floors and cabinets with a solution of one-half cup baking soda in a bucket of warm water. You will have to empty the cabinets before you do this. Change the water in the bucket at least four to five times if you are washing a 1500 square foot home. Sponges and a sponge mop will help.

After each wall is washed, wipe the washed areas with rags that have been laundered in baking soda and not in fragrance detergent. This is time consuming, but it is about as inexpensive as you can get. Baking soda is also available at most markets. You should be able to buy it for pennies a pound.

Enough about VOCs. Let's review the other big indoor problem category: particles.

Particles include anything that can be seen under the microscope such as pollen and mold, plant parts, dust, cat and dog antigen, diesel exhaust and latex from rubber tires

Pollen: An indoor pollen problem can result during the spring if you or your neighbor have a lot of olive and mulberry trees on the property. Only a tiny percentage of pollen gets into the air. These lay on the ground near the plant that produces it. The pollen can enter the house by air currents, but mostly it will enter by tracking on the shoes.

Depending on where you live in the country pollen is also a big problem in the months of July through October when allergenic weeds and grasses are abundant. These enter the home by tracking, as well. The more summer rain we get, the bigger the fall pollen season we can expect. Solution: Remove your shoes at the door.

Mold: Indoor mold damage can occur if there is a burst pipe or roof leak, the swamp cooler is not cleaned every couple of months or other ways in which there is water damage to the residence. Unlike pollen, the water damage must be dealt with immediately or mold in the home will likely result. The damage caused by water is insurable, the mold damage is not.

Plant Parts: Plants give off microscopic particles other than pollen. These particles are also allergenic. Ragweed and juniper are examples. Plant parts are a big part of household dust and also enter the home by tracking on shoes as well as by air currents.

Two things of interest about plant parts: They have many of the same antigens as does the pollen from the same plant, and they are produced at a different time of year from that of the pollen.

Dust: People aren't allergic or sensitive to dust. They react to dozens of things that are in dust. This includes cat and dog antigen, lead, and pesticides. There are more of these indoors than out-of-doors.

Also, in house dust are harmful carbon particles from automotive exhaust and latex from tires. These are allergenic unto themselves and also aggravate allergies. Solution: Keep the home dust free and remove shoes at the door.

Ideally, how do we keep a home dust free? Fortunately, there are simple rules to resolve this issue.

Avoid soft furnishings. These retain dust, and include carpeting, cloth armchairs and sofas, porous drapes and bedspreads. If you have these items then vacuum them slowly and carefully.

For example, carpets are fine if you follow manufacturers recommendations. Most people don't. They vacuum quickly with a vacuum cleaner that has poor suction. Both are no-nos.

Keep the master bedroom as simple as a motel room and throw the bedspread in the clothes dryer (on "AIR" setting) at least once weekly to shake loose the dust. Along these lines, avoid the use of fragrance products in the washer and dryer. You have to spend hours each night breathing in the perfume.

Go through the home room by room and decide how to simplify it. Remove items that lead to clutter.

Make sure you dust items above eye level.

Outside you may want to plant a stand of oleanders as a windbreak. A fence with gaps between the slats is better than a solid fence for filtering out dirt. Wind tends to flow over a solid object such as a home and bring the dirt into your back door.

If you buy an air purifier, make sure the carbon filter is on the inside not the outside. Otherwise grease and dust that is in normal air will block the pores in the carbon and it will cease to absorb odors after a short period of time. Buy a unit that can go into the master bedroom or child's room. Make sure it is rugged and has a good guarantee. Factor in the cost of replacement filters when you purchase the unit.

Beware of claims that say their particular brand will make you feel better. That's called false advertising and the federal government frowns on that. If you hear of such a claim ask to see evidence that it works, not testimonials.

At best, a unit will remove airborne dust. It will not remove the dust from your bed covering, from under the bed, in the carpets, from your sofa, from the window sills, from your TV screen or from the grout between your tiles. You will have to do that part. There is no magical cure.

FIXING THE DUST PROBLEM

While there are 3,000 cases of cancer each year due to outdoor pollutants, there are 20,000 cases due to indoor pollutants. There is no question but that quality of life can improve with indoor air cleanliness, according to studies.

The dust and all that is in it found in old carpets, sofas, and mattresses appears to be a major source of toxic exposure to us. They can have 400 times the level of carcinogens and lead compared with bare floors, window sills and other bare surfaces. This is because they have not been maintained or cleaned adequately over the years. Surfaces are another problem since it was found that there was lead in almost all dust samples taken from window sills and other surfaces. The more dust, the more lead.

Another indoor pollutant includes chloroform from hot showers, clothes and dishwashers. Chloroform is a probable human carcinogen.

So where do we go from here?

Air the home on a regular basis (not in pollen or mold season if you are sensitive to these allergens) or use a well maintained swamp cooler, or both.

Use a good quality air filter, such as a pleated filter, in your central heat or air conditioner system. This includes the use of a good quality air purifier for the master bedroom. It has been recommended that regular damp dusting of surfaces and use of a HEPA filter vacuum cleaner on rugs and carpets will help. I recommend that we avoid the use of products such as fragrances and perfumes. These are from an unregulated industry. Many of these are known respiratory irritants and carcinogens.

Here are other solutions:

1. This is probably the most important recommendation of all. Remove your shoes upon entering the home: this means removing them as you enter indoors. This will decrease the amount of tracking indoors of toxics by a factor of five to ten. Even a front and back scrub mat will make a big difference.

2. Use an efficient vacuum cleaner with a power head once a week on rugs and floors (twice a week with a crawling child in the home).

3. Once a month, vacuum and/or wet-wash surfaces that may be touched (furniture, windowsills, children's

toys, and car interiors). Hand vacuum with a power head on plush upholstery.

4. Choose furniture, floor coverings, and curtains that are easy to clean. Bare floors, flat rugs, and flat upholstery are easier to clean than plush upholstery.

5. Clean carpets annually with a truck-mounted hot water extraction system.

6. Clean air ducts every two to three years.

7. Use the least-toxic products available for cleaning and other home uses.

8. Use your bathroom and kitchen exhaust fans.

9. Vacuum area rugs on both sides since a single pass on one side will only remove 5-15 percent of the dirt. And it is the dirt we are after.

10. Air dry-cleaned clothes outside for at least a day before wearing them. Realize, however, that unless they are aired in warm to hot weather the perchloroethylene will not evaporate.

Household dust is not all that harmless. It is made up of dozens or more allergens and irritants. These include pollen, mold, pesticides, foods, insects, latex and dirt.

There are a lot of advantages to organizing the home for reasons of health. One nice thing is that it can be done inexpensively. In the end you will have more free time because cleaning will be easier. While it may not be fun, it provides the excuse to throw away, give away or store items that you no longer need or want.

So, papers lying around, books and videotapes that should be behind cabinet doors, and a mess in general is part of our lives. It doesn't have to be. Experts agree, if dust and allergies is your problem do yourself a favor and reorganize your home.

The worst areas of the home are as follows:

1. Counter tops: Kitchen appliances should be put back in cabinets after they have been cleaned.

2. Areas above eye level: This includes the top of the refrigerator, pictures, window ledges and ceiling fan blades.

3. Desk and work area: Part of our clutter problem is that we just don't know where to put things. If the philosophy of "When in doubt throw it out" is too drastic for you then create more files. The office area is like any other area. It can be a runaway problem.

4. Books and bookcases: Not only the shelves but the tops of the books retain dust. See if you can find some way to enclose those books. Simplify your home and the master bedroom, as well.

5. Decorative items: These are the toughest to clean because they have so many grooves and dust-settling areas. They are so personal that we hate to put them into storage. However, pictures on walls retain less dust than do pictures standing up and items behind glass gather little dust. Locate appropriate cabinets to display your items behind glass.

6. Closets: Nobody ever dusts them but there are numerous ways to reorganize them so that you can. Closet reorganizers can be purchased or you can do your own. Throw out the old. Spend time in those closets. Once the closets have been totally reorganized they will be like new additions to the home. Damp dust only in here, unless you are going to take everything out and start over.

NON-TOXIC CLEANERS

One may reduce daily personal chemical exposures by selecting soaps without artificial color or fragrance. A non-toxic shampoo can be made with one part olive or avocado oil, two parts distilled water and four parts Castile soap. Baking soda and cornstarch in equal amounts create a non-toxic deodorant powder for adults.

Herbs such as onions and mint, can be planted around buildings to discourage entry. of insects. Boric acid powder is a safer alternative to sprays for ants and roaches. Slugs and snails are easy victims for beer traps. For pets, toxic flea collars can be replaced with herbal collars and ointments made from eucalyptus or rosemary. Brewer's yeast in an animal's diet also discourages fleas and ticks. Instead of carcinogenic mothballs, try cedar chips or lavender flowers. A solution of bar soap and water can be sprayed on the leaves of houseplants to kill pests, then rinsed off.

An all purpose cleaner can be made by combining a teaspoon each of liquid soap and borax in a quart of warm water and adding a squeeze of lemon. Half a cup of borax to a gallon of water makes a household disinfectant.

Linoleum floors are effectively cleaned with half a cup of white vinegar to a gallon of water. Baking soda substitutes for scouring powder and removes porcelain stains. Wine and coffee cup stains can be removed with moist salt. Carpet upholstery stains may be removed with club soda. Rust spots on clothing can be bleached out with lemon juice and sunlight. Spills in the oven should be sprinkled with salt immediately; then moistened and brushed with baking soda after the oven cools.

If you noticed, lemon is mentioned a couple of times in the preceding advice. Read on to discover the real magic of lemons.

We have been led to believe that hazardous commercial products are best for carrying out household chores. Certainly, in their own way, they do the job. But our use of these products is only a conditioned habit. As such it can be broken. Take a look at the use of an inexpensive product that will not only lighten the load on your respiratory tract but save money and not take up much space under the sink—the lemon.

Shower doors, tiles and other glassware: Water stains are left by hard water that contains lime. This is chemically basic in nature. A dilute solution of warm acidic lemon juice will quickly clean water drops from bathroom tubs, doors and fixtures. Half a lemon squeezed into a quart of warm water should do the trick. Use a sponge dipped in the juice and wipe down. Then dry. Some people use vinegar to do the same job. I prefer to use lemon juice because it has a more pleasant smell.

Faucets: Avoid toxic commercial polishes. Rub with lemon peel, wash and dry with soft cloth to shine and remove spots. This natural method of cleaning has disinfectant qualities and leaves the faucets with a fresh smell.

Odors: A half lemon placed on the shelf of refrigerator will absorb odors.

Tubs and sinks: Instead of using heavy cleansers with bleach and fragrance, rub with half a lemon dipped in borax. The lemon will take care of the lime build-up and the borax will take care of the soap and grease.

Copper and brass polish: Commercial polishes and cleaners are toxic to inhale. If they can be smelled then they can be asthmagenic and act as a respiratory irritant. Instead, use lemon juice or a paste of juice and salt. Rub, wash with distilled water, dry.

Furniture polish: Mix one part lemon juice with one part olive oil or vegetable oil. You may need to increase the amount of oil to your preference but the furniture Is left with a fine gloss. No vapors of petroleum distillates pollute your airways.

Humidifying devices: use lemon juice solution as a descaler.

Paint brush: Instead of soaking in toxic paint remover, dip hardened paint brushes into boiling lemon juice.

Lower heat immediately. Leave brush for 15 minutes, then wash in soapy water.

Paint: A few drops of lemon juice in outdoor house paint will keep insects away while you are painting.

Paint on glass: To remove dried paint apply hot lemon juice with soft cloth or paper towel. Leave until nearly dry, then wipe off the old paint.

Garbage disposal: Toss used lemons into your garbage disposal to help keep it clean and fresh-smelling.

Note: As an alternative to lemons concentrated lemon juice is available in major markets and is very inexpensive.

HOME STORAGE OF CLEANING AGENTS

"Solvent" is a very broad term encompassing a wide range of liquids that are capable of dissolving or dispersing other substances. They are found in many products commonly used at home and at work, for example, in paint, varnishes, adhesives, pesticides, and cleaning solutions and are among the chemicals most frequently implicated by chemically sensitive (CS) patients. The volatile organic compounds associated with sick building syndrome are in large part solvent vapors.

There is a curious range of effects caused by solvents, according to Ashford and Miller (See References). These include effects upon normal individuals such as being alert, enthusiastic, energetic and witty, tenseness, behaving jittery, and argumentative. There is also a withdrawal phase whereby normal persons report allergy-like symptoms and CS persons report fatigue, depression, headache and joint aches.

If you are seeking escape from solvent vapors be aware that there must be plenty of ventilation to dilute the chemical gas. This means getting plenty of fresh air during and after painting (as well as seeking a paint that uses a water-base, and taking plenty of rest breaks) and deciding which is the upwind side of the hose you are going to use while pumping gas.

Virtually every home has a large number of solvents, paints, varnishes, detergents, household cleaning products, aerosols and similar products stored under the sinks, in the laundry room cabinets, garage and in various locations. This means that household members, guests and pets are constantly exposed to various levels of potentially harmful substances.

Ensure that the tops of all containers are well closed and wash or wipe spills that may have occurred on the side of the container or on the shelf. Also, obtain a plastic container of appropriate size in which these substances can be stored. Make sure you get a tight fitting top for it If you are going to go this route, then wipe down the shelves first to remove the odor of chemicals that has layered onto them. Use a damp wash cloth and baking soda, but try to avoid the use of chlorine bleach because this is a respiratory irritant.

NEWSPRINT BREAKTHROUGH

Over the past several years (if not decades) there have been a number of rumors, personal reports, word of mouth testimonies and medical cautions regarding the potential hazards of newsprint ink. It seems as if everybody has an opinion about this subject. A cause and effect relationship between this ink and symptoms in persons sensitive to (VOCs) has been difficult to prove. In fact, even sensitivity to volatile organic compounds (VOCs) in general is met by skepticism by much of the medical community.

Terms like "hog wash" are not uncommon when some doctors are asked about chemical sensitivity as a problem afflicting society. But the matter brought forth the questions: Just exactly what is in newsprint and where is the industry headed?

Historically, inks were comprised of chrome, lead and flammable solvents. Paper dust was as much a problem to the newspaper recipient as it was to the newspaper worker. Think about reading a paper and scratching your nose or touching your eyes with that stuff on your hands and in your lungs.

In 1989, President Bush proposed legislation that would update and strengthen The Clean Air Act that was passed in 1970. A major portion of that bill is directed toward decreasing VOCs that are found in hydrocarbon based petroleum inks.

In this regard, for several years research has been ongoing with the hopes of completely switching to the use of soybean based inks (SI). It has been found that SIs have several advantages over petroleum based inks (PI). For one thing, SIs do not emit VOCs at nearly the extent of PIs, they are environmentally safer, there is less ink transfer to the hands (especially when no-rub inks become universally adapted), soy is grown in the United States (Midwest), and the price is comparable when colored ink is used. Importantly, much of the ink is recycled.

Most newspaper inks use the same soy oil that is used in cooking. Add to it a small amount of coloring agents and binder (resins) and there you have a product competitive to petroleum based ink. In fact, at the end of the first marketing year, according to the American Soybean Association, six newspapers were using SI. One thousand newspapers were using the ink at the end of the second year. After three years of availability, soy ink was being used by one-third of the nation's 9100 newspapers, including one-half of the daily 1700 U.S. Daily newspapers. A number of Midwestern states have passed laws mandating the us of SI for state government printing.

When newspapers convert from petroleum base to SI, it is due to a number of factors: environmental safety, ready availability, less exposure to VOCs by staff and the public, better recyclability of newsprint, better color, and pricing competitive with PI. This latter is important since a newspaper with a daily circulation of 75,000 spends approximately $700,000 yearly for newsprint ink.

Admittedly, there is a catch. As noted by virtually all parties involved, SI works best for colored newsprint, fliers, brochures, catalogues and other printed products. The drawback comes when black ink is involved. The reason is cost. Surprisingly, it costs almost twice as much to print in black as it does to print in color per unit. This is because of the difference in price between soy oil (the ink solvent) and petroleum oil.

In the case of colored soy oil ink the price depends upon the cost of the colored pigment and not the price of the oil. This is because even normal colored ink is expensive and so is the soy oil solvent. Thus, the printing industry might as well go to soy.

With black ink, the real cost is due to the cost of the solvent (soy oil) and not the ink pigment. It is cheaper to use petroleum based solvents. Petroleum based oil is a lesser grade of oil than that used for colored ink and it costs one-half to one-third as much as does soy. (Black ink itself is cheap since carbon-black is made from the burning of petroleum.) If it cost the same, then the price would be passed on to the consumer.

According to the American Soybean Association, about 100 million bushels of soybeans are used per year to manufacture all kinds of publishing ink; 30 million of which are used for soy-based newspaper ink. Use of soy oil is a function of supply and demand, and as more soy is used for oil purposes, the cost will come down. From a newspaper publisher's standpoint, soy oil works about as well as does petroleum based solvents. It is only a matter of time when economics will permit the universal use of SI.

The rub comes here: Since the Department of Agriculture limits the amount of soy that is grown in the United States the price of soy stays high as opposed to open market economy prices. This forces the print media to stay with petroleum based oils.

Regarding black ink, the newspaper industry has no qualms about switching to soy based oils, given reasonable costs of this oil.

In summary, new no-rub inks and new petroleum oils that have a low content of petroleum based volatile organic compounds (VOCs, lacking in carcinogens) help the industry to maintain a quality black ink for its printing purposes. When the price of soy oil becomes competitive with the price of petroleum oil then the industry will switch to soy oil, which has virtually no VOCs.

CREATE A CLEAN-ROOM

Dust has perhaps fifty or more different particles and gases that make it what it is. These include cat antigen, plant parts, pesticides, laundry detergent, food particles, bugs, pollen and mold, and auto exhaust particles.

Soft furnishings and dust go together. The bedroom has some of both. Add to this length of time we are exposed to the dust and our own sensitivities and we have a bad mixture. Our goal is to control the level of dust in the home and especially in the bedroom where we spend so much time. If we can do this, we will have minimal exposure to allergens and thus be better able to deal with the allergens outside the home. There is a certain pleasure in knowing that there is at least one area of the home that is allergen free. In other words, it is the clean room.

When we can create a clean room, then our reaction to the environment in general will improve. Have some fun and reorganize.

The clean room needs to be as simple and plain as possible with a minimum of soft furnishings that include arm chairs, lace curtains, carpeting, and porous bedspreads. Use simple bedspreads that can be thrown in the clothes dryer set on "Air" to remove the dust. Dacron, polyester or other synthetic fibers can be used for pad fillings as well as for pillows. Never use feather, kapok, foam rubber, or cotton pillows since these can break down in time and either leak particles into the air or become contaminated.

Use washable plain cotton curtains in the room. Drapes requiring dry cleaning should not be used, since the commercial dry cleaning fluids out-gas over a long period of time. Avoid any shag or soft velvety fabrics for curtains. Do not use venetian blinds unless they are vertical. Keep it simple.

No shoes allowed in the bedroom. If you have an air purifier, keep it running at night. You will soon get used to the background noise, which will override other sounds of the night.

Weather-strip the windows to minimize the influx of dust into the room.

The less furniture in the room the better it is for your respiratory health. Furniture should be simple, not ornate.

Clean out the closet and only keep the most frequently worn clothes in it. This is because clothes retain dust. They also emit odors from fragrance detergent and fabric softeners or odors from new clothing. The idea is to minimize your exposure to airborne irritants, whether they are particles or gases. This means that you will have to find space somewhere else in the home to put the clothing that is not worn quite so often. Winter clothes should be bagged. Do not permit clothing or papers to carelessly lie about.

Most people pick the master bedroom for their clean room. Keep this room as simple as possible. Make a game out of changing the lifestyle a little bit. After all, it is a free way to reduce your health risk.

Wall hangings should be minimized since dust will settle on the top surface above eye level.

Discourage the use of bookcases for the same reason and minimize the use of ornaments and small items that generate more dust than pleasure.

The use of a thin layer of cheesecloth behind air registers helps reduce dust entry into the bedroom. This must be changed occasionally to prevent reduction in airflow.

Do not use insect sprays or powders in the bedroom. Many of them contain allergens or chemical irritants. Avoid the use of deodorizers and mothballs since virtually all of them have been shown to be carcinogenic.

Blankets and clothing that have been stored for several months should be thoroughly aired out-of-doors or tumbled in the clothes dryer before being used.

Remember that the master bedroom is usually connected to the master bath and the master bathroom can be a real trouble spot, unless it is well maintained. This is because it contains numerous fragrance products such as toilet paper, facial tissue, soaps, cosmetics, shampoos, conditioners, hair sprays, perfumes and deodorants. Also present are various cleaning agents, polishes and solvents. The odors from all of these communicate directly into the master bedroom and of course, you get the full dose while in the bathroom.

What to do here? Just keep the bare minimum of supplies in the master bathroom and move the rest somewhere else. Scrub out the storage shelves with a solution of baking soda (one-half cup to a bucket of warm water) to absorb the leftover odors. And use the fan frequently. That is what it's for.

There is mounting evidence that house plants are beneficial in terms of removing odors and dust from home air. However, most of the reports recommend against plants in the home environment of a patient who is dust sensitive. This is because dust gathers on their leaves. Others say that's the whole point. If you must have plants in the bedroom, the state of health of the plants is probably more important than any other factor, since diseased plants liberate mold spores into the air.

Artificial decorative plants are out since they will also accumulate dust.

Keep the bedroom dust free with frequent cleaning. Wipe floors, shelves, tables, and other furniture with a damp cloth oiled with a mixture of one part lemon juice and one or two parts cooking oil or olive oil.

Assuming your ducts are not leaking you should now have a dust-free and smell-free room or two. So no matter what else is going on outside as far as wind, pollen, mold or irritants, this room will be a safe haven for you.

These are all proven methods. They are not extreme and are recommended by allergists and researchers around the world. This is an important subject because it is something we can all do with minimal spending.

RESPIRATORY PROBLEMS AND INDOOR HEAT

What would you say if the weatherman forecasted moderate to cool temperatures over the next several days, weeks or even months with the relative humidity at or below twenty percent, and even lower in some cases? Should you be concerned?

Second, do you have the ailment that is going around? The symptoms are dry and scratchy throat, sometimes to the point of soreness in the morning, itching eyes, scalp and skin, and do you talk like a frog when you wake up?

Are your allergies activated outdoors and indoors even though the usual allergens are at a low level outside and you can't think of any way that your indoor lifestyle has changed?

Finally, if you don't have allergies are you still having problems? Welcome to the world of winter with the heater turned on.

Depending on the amount of draft and leakage in your home, the level of indoor humidity will vary, but expect the vast majority of homes and apartments and other dwellings and businesses to experience a crash in relative humidity during the cool evening hours. Space heaters create the least problems, central forced air creates the most problems. The higher the heat and the tighter the home, the worse the problem becomes.

Some scientists believe that dry air is a major factor in causing us to have more symptoms of allergy, asthma and other respiratory disorders. They believe that the respiratory tract becomes primed when exposed to a low level of moisture for a period of time. This certainly fits right in with what is known about asthma.

So we go home (or to school or to work), turn on the heater and burn off the moisture. We expose ourselves to this environment all evening. Senior citizens and shut-ins could be exposed to extremely dry conditions for days or weeks on end if not longer. In the world of indoor air quality this is called being out of the comfort zone.

Why are people reacting as if it is pollen season already? This is because you don't necessarily have to be exposed to a high dose of allergens to have symptoms under these dry conditions; the same old ones will do just fine. These allergens include dust, cats, dogs and the usual assortment of allergens that trigger symptoms.

What to do? Be aware that you are not catching some dreaded disease. A glass of water next to the night stand and periodic sips during the night will help the throat and add needed moisture to a dehydrated respiratory tract. Hair spray dries the skin; moisturizer helps. Also, to add moisture to the air, don't run your range or bathroom fans during cooking or showering.

One thing that is common to virtually all homes: a central heat supply. Early thought needs to be given to turning on that central furnace. Let's take a close look at why the unit frequently stinks when it is first turned on and why it causes so much respiratory distress.

There are a number of reasons for this problem. First, the heater unit is no different than any other area of the home in that it accumulates dust and worn fibers. The difference is that it doesn't get dusted. The fibers come from natural and synthetic materials in carpets, furniture, draperies and clothing. They include natural fibers such as cotton and wool and synthetics such as nylon, rayon, polyester and numerous other varieties.

The gases released are the same or similar to those released during a fire. When these fibers are heated or burned they give off small amounts of annoying, if not toxic gases. These gases may include hydrogen cyanide and acrolein.

Hydrogen cyanide is released when rayon, nylon and other synthetic fabrics are burned. Acrolein is released when plant particles are burned. They are respiratory irritants. Also, aerosols that we use during the year will find their way to the heating unit and the duct system to become warmed, heated and sent back into the home. These aerosols will include pesticides, herbicides, furniture polishes and cleaning agents.

In addition, there can be a certain amount of condensate in the ducts and furnace where microorganisms may flourish. They can get burned off, too. And remember that basic smells of any home are a lot more noticeable when they are warmed.

Gas furnaces have an additional problem in that there is a distinctive odorant (mercaptan) added to the gas so that you can smell it and will not confuse it with another smell, in case of a leak. Uneven heat at startup leads to uneven burning and the odorant is a lot more noticeable than at subsequent startups.

Many persons have become ill at the smell of first heat. Some have not used the heater for the remainder of the winter thinking that the smell was permanent. Be kind to yourself. Ventilate. It may be cold outside, but open the windows a little, then turn on the heat. It shouldn't take long for the smells to dissipate. Chances are you will now have a warm home that has little or no smell for the rest of the season.

As a final note, if you have had a gas furnace for a number of years, then you need to be aware that there may be hairline cracks in the heat exchanger and deadly carbon monoxide gas could be entering your home. Call a your local gas company for a licensed contractor to check out the furnace for leaks.

SMELLS IN NEW CARPETING

What causes the smell in new carpeting? The backing on a carpet is rubberized and a chemical known as 4-PC used in the backing can be smelled at very low levels. Synthetic carpets will have more odor than natural fiber carpets. It has been some fifteen years since formaldehyde has been used in domestic carpets.

A carpet consists of face yarns attached to the primary backing of the carpet. This is usually coated with latex rubber to lock the yarn to the backing. Sometimes a secondary backing is added for stability. The rubber is a styrene-butadiene complex that does have some VOC emissions. The best known is 4-phenylcyclohexene (4-PC) which has an odor detectable at an extremely low level. This is the odor that most people associate with new carpeting. Other emissions include styrene, decane, toluene, xylene and a variety of others hydrocarbons.

Commercial carpeting used in offices and schools is typically glued down to a concrete surface that has been prepared with a sealer.

It is the adhesives that emit the high percentage of VOCs, not the carpet itself. Carpet adhesive is a mixture of synthetic rubber, resins, and fillers. When solvents are used to dissolve the resins into a liquid, more than 90 percent of the VOCs are produced at that time. Because of this problem, adhesive manufacturers are developing low-solvent and solvent-free adhesives and seam sealers.

The glues used with commercial carpeting emit most of the volatile organic compounds (VOCs), not the carpet itself. Still, the latex backing in household carpeting releases a significant amount of VOCs into that environment, although not nearly as much as does the painted walls in a new home. There is no scientific information that links VOC emissions from carpets to health effects in humans in general, but certain individuals are reactive. (An odor is not considered a health effect.)

Ideally, the new carpet should be unrolled and aired out for a minimum of 48 hours prior to installation, whether it be in a school, home or business. The building ventilation system should then be run at maximum air and normal temperature for 72-hours after installation. Use of return air supply should be minimized. If possible the home, business or school should be opened as much as possible during this three day time period. Exhaust fans can be placed in strategic locations.

In a building, recirculation of indoor air should be avoided for a number of reasons. For one, the air should be expelled; for another, ceiling tiles, duct liners and other structural materials as well as soft furnishings are capable of absorbing the odors and re-releasing the VOCs.

For home use activated charcoal filters in a polyester matrix are available for the central air supply and return air ducts after the 72-hour time period. Activated charcoal is inexpensive and absorbs many of the VOCs emitted from new carpeting.

In all circumstances carpeting should be vacuumed prior to removal, ideally with a HEPA (high efficiency particulate air)vacuum. Then the floor should be vacuumed and mopped. This will minimize the amount of dust generated during carpet removal and installation. Finally, vacuum the new carpet after installation to remove loose fibers. Canister vacuum cleaners are generally better than the upright variety since fine dust is emitted from the bag of the latter. A slow forward and backward motion over the same spot is better than rapid motions.

In schools or offices, heavy traffic areas should be vacuumed daily while light or medium traffic areas can be vacuumed once or twice weekly, or by inspection.

The home carpet that is merely tacked down has very little in the way of VOC emissions. However, the symptoms mentioned above can still apply. This is due, in part, to VOCs acting in combination.

Over a period of time the EPA and CRI met over this issue. The CRI developed a set of industry standards regarding the level of emissions from styrene, 4-PC and formaldehyde. It has been a dozen or more years since domestic carpeting has contained formaldehyde but foreign made carpet may contain the irritant.

The CRI tests batches of carpet once yearly. This is not frequent enough according to the Consumer Product Safety Commission. The latter found that batches of carpet vary considerably in their chemical composition over that time period. Another problem is that when the carpet meets CRI standards it is issued an indoor air quality label. Be aware that this is not a label of safety for the consumer.

Anderson Labs of Dedham, MA passed air over new carpets. This air caused nervous system disorders in the mice that it didn't kill. However, these experiments have not been duplicated elsewhere.

Another event occurred a few years ago when the Environmental Protection Agency (EPA) laid 20,000 square feet of new carpeting in its eastern headquarters and hundreds of workers complained or became ill.

The Carpet and Rug Institute (CRI) represents the vast majority of carpet manufacturers. Neither the CRI nor EPA links emissions from carpets with specific health effects. The key word here is "specific" since each individual can react differently to the same volatile organic compound (VOC). Despite their denials the CRI is under pressure. This is because of the publicity surrounding the Anderson Lab test results. Also, it is now generally recognized that the most frequent effects of low level exposures to VOCs are stimulation of the senses, inflammation of exposed tissues, and stress-like reactions such as headache and fatigue.

ZERO EMISSION PAINT

The paints of today are of a much better quality than a couple of decades ago. At that time they contained lead and a high level of what was known as white spirits. The waterborne paints of today still have their share of respiratory and skin irritants. Waterborne paints include acrylic latex paint (flat and semigloss), latex enamel, latex wall paint (regular and heavy bodied), latex primer, and sealing waterborne paint. They contain water, biocides, surfactants, white pigments and extender pigments, latex monomer precursors, solvents and cosolvents, driers, plasticizers, amines and various volatile substances. Some of these have a pungent odor.

Medical studies have found that the ingredients in these paints can irritate the skin and mucous membranes and trigger bouts of depression. The volatile substances include ammonia, butyl acrylate, styrene, white spirit and formaldehyde. The painter should always maintain good ventilation when painting is being conducted. Office workers and family members should not have to endure the hazards of breathing chemicals released from paints in a confined space. The paints of today are of a much better quality than a couple of decades ago. At that time they contained lead and a high level of what was known as white spirits. The waterborne paints of today still have their share of respiratory and skin irritants.

Waterborne paints include acrylic latex paint (flat and semigloss), latex enamel, latex wall paint (regular and heavy bodied), latex primer, and sealing waterborne paint.

They contain water, biocides, surfactants, white pigments and extender pigments, latex monomer precursors, solvents and cosolvents, driers, plasticizers, amines and various volatile substances. Some of these have a pungent odor.

Medical studies have shown these paints or their ingredients to irritate the skin and mucous membrane and trigger bouts of depression. The volatile substances include ammonia, butyl acrylate, styrene, white spirit and formaldehyde.

While drying, oil-base paints emit a very high level of toxic volatile organic compounds (VOCs); much more than wall coverings, wood products and carpets put together. Paints are a major factor in the irritating smell of a new home, much more so than the carpeting. Often, they contribute to a great deal of respiratory illness, headaches, nausea, dizziness and other symptoms in the average residence. If a newly painted home is closed up during hot sweaty weather, then the smell absorbs into the walls and carpets (and furnishings, if present) and will last for a weeks or months. In fact, the VOCs emitted from paints far exceeds that of new carpeting.

The VOCs found in many paints include such toxics as xylene, styrene, benzene, and decane. Read the Material Safety Data Sheet available at the paint store for the average paint for more information.

Wouldn't it be nice to have just one builder declare that he has built a tract of homes with zero emission paints? One problem with low emission paints is that they lack good opacity. Therefore they can't cover very well. Plan on using a couple of coats.

Some paint manufacturers make low emission paints and many other companies are coming into the fold. So ask any paint dealer if they carry the right product. There may be some available in your area.

THE OFFICE ENVIRONMENT

Some of the office problems include lack of fresh air, cigarette smoke, paints, carpeting, cleaners, perfumes, plastics, formaldehyde, pesticides, ozone and plastics from electronic equipment. Your own experiences may allow you to add some names to this list.

Carbonless copy paper is made with micro-encapsulated color formers on the back of the sheets and color formers on the front of the paper sheets. When pressure is applied by a ball point pen the dyes in the capsules are dissolved in various oils and other chemicals. These chemicals include phenyls, hydrocarbons, ethanes, naphthalene, chlorinated paraffins (waxes) and benzenes.

The developers most commonly used in the United States are phenolic resins, that is, hardeners related to benzene.

This chemical soup is a mixture of bad guys. To date there have been no conclusive studies linking health effects in the office with the use of this paper. However, its use in poorly ventilated offices should be discontinued. This is on the grounds that the chemicals released through its use only add to the chemicals in the office air. The end result is that the amount and quality of work falls off.

Typewriter correction fluid is a major source of volatile chemicals in the office building and in the home office. It may contain trichloroethylene, cresol, ethanol and naphthalene. All of these chemicals are toxic when used alone and much more toxic when inhaled together. Look for water-base correction fluid. You may have to use two coats.

Become familiar with the names of these toxic chemicals. Then you will learn to avoid them.

Finally, if you have to go outside the building to smoke make sure the fresh air vent for the building isn't located over your head.

One of the most common problems in the office environment is the copy room. This room usually consists of numerous machines in a confined space. Several persons are frequently using the machines and ventilation in the area is poor. Ventilation here is the key to fresher air. Fresh air dilutes machine by-products that can be irritating to many people. Ozone is one example of such an irritant in this setting. Mimeograph fluid is another.

Another problem is an extremely high level of airborne and settled dust. Since numerous persons are usually present in the typical office, there is a significant amount of tracking of dirt indoors on shoes. This is ground to a fine powder that either settles into the carpeting or becomes airborne. To minimize this problem, a conscientious cleaning staff and a good quality vacuum cleaner will help reduce the indoor particle load.

Also, check the cleaning supplies. Ensure that they are as odor free as possible.

HAZARDS OF COSMETOLOGY

An important, if not necessary part of our daily lives is hair and beauty salons and their supply stores. The beauty industry is worth over a billion dollars to our national economy. The beauty business has not only played a part in the lives of women since recorded time, but is becoming a part of the life of men as well. To accept these salons is to accept their odors. The chemicals that cause these odors have come under scrutiny over the past decade or so. Where once we accepted these odors as almost natural we now examine them from the standpoint of health, not only to the people who handle and breathe them but to the public at large.

Numerous medical reports have appeared over the years regarding the hazards of chemicals used in cosmetology in such respected journals as the Proceedings of the National Academy of Sciences, Mutagen Research, and the journal of Contact Dermatitis. A recent issue of the Journal of Toxicology and Environmental Health contains an article about mutagens found in the urine of cosmetologists and dental personnel. Mutagens are substances that cause mutations, that is, change the structure of DNA.

In the workplace cosmetologists are exposed to a wide range of chemicals, both by skin contact and by inhalation, according to the article. Some two thousand chemicals are present including those in soaps, foaming detergents, shampoos, rinses, conditioners, dyes, bleaches, estrogens, vitamins, suntan products, deodorants, depilatories and hair straighteners, nail builders (hardeners, enamels), propellants and preservatives.

The chemicals include azo dyes, formaldehyde, heavy metal oxides and sulfides, etc. Many of these chemicals are not mutagens as such but are changed to mutagens in the body. In particular, many chemicals in hair dye are mutagenic and some cause cancer in laboratory mammals. Certainly, skin disorders and allergies are commonplace and are well documented among cosmetologists.

Thousands of pages have been written related to the subject of cosmetologists' and dental workers' exposure to toxic chemicals. This says nothing about beauty supply stores. However, in terms of mutagenic effect and cancer it must be stressed that the long term effects of exposure to these chemicals by the workers in these work surroundings is a big unknown.

According to an article in the Journal of Toxicology and Environmental Health, cosmetologists were at two times the risk of exposure to direct-acting mutagens compared with dental workers.

To their credit many hair salons are now using carbon filtration air purifiers to remove the odors from the work place. Development of multiple chemical hypersensitivity (MCS) by employees and the public at large may be another story.

ART AND CRAFTS

Question: What business sells the following products? acetone, methylethylketone, toluene, 1,1 trichloroethane, aliphatic hydrocarbons, xylene, hexane isomers, propane, diacetone alcohol, turpentine, linseed oil, d-limonene, wood stain, spray enamel, epoxy paint and adhesive caulk.

Answer: Art supply stores. These stores sell graphic and art supplies to amateur and professional painters, architects, schools and parks and recreation art students.

Historically, artists as painters have encountered toxic heavy metals such as cadmium, iron, lead, silver and gold, toxic aerosols, and in more recent times, volatile organic compounds (VOCs) made from petroleum products. Many of these VOCs are listed above. Nowadays, most art supply stores are health conscious and environmentally attuned. They have no choice. Like almost every other business, art and its relatives are becoming health conscious.

For a long time the arts and crafts business/industry has been loaded with poisonous materials. These ranged from heavy metals such as lead, gold, silver and cadmium found in dyes and stains to oils and solvents that are inhaled. In recent years the industry has responded to social pressure and common sense and it has gone to great lengths to create non-toxic water-base materials.

The xylene-free non-toxic marker is but one of many examples. Since xylene, toluene and benzene are closely related chemicals and are all toxic the creation of these new markers is progress for cleaner air in classrooms and meeting rooms, for children at home and for the rest of us who have used them from time to time on an individual basis.

More solvents are going away from a petroleum base toward more water based including water based acrylics. One important example is markers. At one time markers were noted for their ability to fry brain cells because of their toxic components. This became a real problem with children and teachers. Now, many markers are labelled as non-toxic and do not give off toxic vapors. Pigments used for water colors and oil painting contain less heavy metals with each passing year just as home paints have eliminated lead and are generally safer than a decade ago. Another example is canned air used for air brushing, which now uses an environmentally safe propellant. Still in all, maintaining plenty of ventilation while engaged in the practice of painting cannot be overemphasized.

Many high school and college courses still use traditional materials that contain xylene, toluene, carbon tetrachloride, trichloroethylene and other hazardous substances. Additional hazards are present in these courses for the students and instructors. A report in the Journal of Environmental Pathology, Toxicology and Oncology discusses these problems.

This discussion also applies to home workshops, summer programs, artists, house painters and a variety of hobbyists.

First is the problem of non-awareness, otherwise known as lack of education. The instructor needs to learn about potential and real problems in order to pass on the information to the students. Complacency results when non-awareness happens with familiar items.

In our example with markers, the deliberate and purposeful smelling of xylene can become habit forming and very dangerous to brain and lung function. It's in the same league as glue sniffing. Even constant exposure to xylene at low doses can be hazardous over time. If we don't know about the danger of xylene then we drop our guard and get hurt. This is because xylene, toluene and related chemicals are soluble in fatty tissue such as is present in the myelin covering of the central nervous system.

Depending on what we are complacent about we develop overexposure and typical symptoms of sneezing, itching and watery eyes, runny nose, blocked nose, coughing spells, headaches, itchy skin, redness of skin, shortness of breath, dizziness, memory loss, fatigue and neurological disorders.

The next problem after non-awareness is poor ventilation of the work areas. Typically, room windows are closed because of air-conditioning, if it is on at all. (A room temperature that is uncomfortably high will increase the absorption of solvents through the skin and mouth.) Entrance doors are often closed because of privacy.

There is no special ventilation to accommodate the two major potential hazards: solvent fumes and dust that inevitably result from the work. This is a worse case scenario that happens too frequently. Did you ever paint a room or walk into a freshly painted room at home with the doors and windows closed and stay there awhile? Fans, open windows and evaporative cooling help remove the vapors. But this is only a quick fix and is no substitute for awareness.

If you have respiratory problems or just want to maintain your hobby in the field of art as safely as possible, then ask your art supplier about safe alternatives to the paints, pigments and sprays.

Most of the exposure to solvent vapors comes through respiratory intake. Secondarily, the vapors are absorbed by the skin and mouth. Spills are a different, but related problem.

Solvent spills are not uncommon in the classroom or at home. Let's take good old paint thinner, carbon tetrachloride as an example here. The U.S. National Institute for Occupational Safety and Health describes this solvent as a colorless liquid with an ether-like odor. Over-exposure can lead to central nervous system depression, nausea, vomiting, liver and kidney damage and cancer.

There is a correct and an incorrect way to clean up spills. Lack of training in spill mopping and proper disposal of contaminated materials leads to further evaporation of the spilled solvent from the table or floor. As we mop the spill the wet paper or cloth frequently comes into contact with the hands. The solvent is absorbed into the skin, blood, muscles and nerves. Thus, absorbent materials should be used for spill mopping held with non-absorbent material to minimize chances that there will be contact of the solvent with the hands. Gloves (non-rubber preferably) will help here. Closed foot-operated cans with plastic liners are preferable to open trash cans for disposal purposes.

Studies have shown that when liquids are spilled there is as much as ninety-eight percent to as little as fifteen percent removal rate of the spilled material. Therefore, one can expect an excessive quantity of harmful vapors to normally enter the air as a result of the spill.

Small particles generated through carpentry are another potential hazard in arts and crafts classes as well as at home. Wood dust, metallic dust and clay generated through wood and pottery work can reach very high levels indoors where there is little or no escape. Masks are usually not worn, nor is ear protection. The noise of power sanders and drills could reach 115 decibels near the ear. If the tools are older, the noise level could be higher. This is at the level of federal guidelines for worker safety.

Finally, we must address the issue of overall neatness. This includes the proper storage and labeling of arts and crafts materials such as solvents, cleaners and finishes. This will go a long way toward reducing the level of vapors in storage lockers and our exposure to them when they are first opened. It will reduce the level of spillages as well as unnecessary skin contact with solvents.

CHOOSING A PEST CONTROL COMPANY

The EPA publishes a guide entitled, Citizens Guide to Pesticides. This booklet does not tell us about pesticides per se', but it does tell us about how to choose a pest control company. At the outset, those of you who have your indoor environment sprayed once monthly and believe that you are chemically sensitive should consider the following two alternatives: cease and desist or desist and cease. There are many non toxic methods to control pests (please see References).

Certainly, termites, roaches, mites, fleas, crickets and ants need to be controlled. This does not mean overkill. There are many reputable pest control companies out there, but like everything else, moderation may be better than the extreme. If you've decided that you need a pest control company to aid you here are some questions you can ask, according to this booklet.

1. Does the company have a good track record? Don't rely on the company salesman to answer the question; research the answer yourself. Ask your friends and neighbors and check with the local consumer office to see if there have been any complaints. One consumer stated that her doctor (a nationally recognized asthma specialist) told her not to have her home sprayed. The doctor's rationale: the sprays contained pyrethrum (related to ragweed) and

volatile organic compounds. The salesperson told her that the doctor did not know what he was talking about; this spray was harmless. While this experience is a rare one, it points to the fact that you should ask questions and ask your doctor if your are really concerned.

2. Can the salesman prove that the company is insured? Contractor's general liability insurance, including insurance for sudden and accidental pollution, gives you as a homeowner a certain degree of protection should an accident occur while pesticides are being applied in your home. Contractor's workmen's compensation can also protect you should an employee of the contractor be injured in your home.

3. Is the company licensed? If it is then there is a certified pesticide applicator present in their office to supervise the work. Make sure the license is current.

4. Does the company stand behind its work and can you stand behind your part of the bargain, e.g., in the case of termite control, a guarantee may be invalidated if structural alterations are made without prior notice to the pest control company.

5. This is most important. Is the company willing to discuss the treatment proposed for your home (and the safety of their product), including special instructions you should follow to reduce your exposure to the pesticide.

Remember that most pest control companies are licensed and reputable. However, the chemicals they use are toxic to life. Frequently, the "consumer" is incompletely informed or does not ask enough questions. You may be consuming more than you know. If there are alternative methods for pest control don't be afraid to say no. And check the alternatives. Many companies will have more than one pesticide on hand. Ask to see the Material Safety Data Sheet on each one and compare them.

THE HOLIDAY SEASON

The holiday season has all of the elements of life's happiness: giving and receiving, friendship, sharing, song and merriment, get-togethers, and lots of food. However, it is also a time when many persons become depressed or ill for days, weeks or even months. We tend to blame these problems on financial pressures, a variety of frustrations and other reasons that are mentally oriented. But consider the depression and illness from a physical standpoint—respiratory toxicity and overdosing.

Overdosing the respiratory tract with airborne chemical substances can lead to depression, behavioral changes and mood swings and can affect the nervous and muscular systems. Entire scientific journals are devoted to this subject area. And our exposure to these chemicals occurs over a short intense period of time during the holiday season. Some scientists believe that this exposure tends to make us more sensitive to indoor airborne allergens as well as food allergens.

Common problem areas:

Fireplaces are known trouble spots because of the outdoor and indoor pollution they create. Many woods contain resins and oils and synthetic logs contain waxes and aldehydes which, when burned, cause significant respiratory and systemic problems from the standpoint of particles and gasses released into the air.

Pine trees are not reputed to have an allergenic pollen, but the smell of pine affects a large percentage of the population.

Cleaners and aerosols are widely used to clean the homes this season during the holiday season.

Fragrances in department stores include those sold and those worn and extra time shopping puts us in closer contact with fragrances in the marketplace and in the homes of others.

Increased activity on the part of household pets permits extra shedding of allergenic hair, feathers, dander, saliva and skin oils.

Increased traffic congestion is the rule of thumb.

Colder weather and tighter homes magnify the effects.

There is increased contact with the homes of friends and relatives where there are non-fragrance gasses from plastics, indoor heating systems, build-up of indoor carbon monoxide and carbon dioxide and off-gassing from furnishings and newer construction materials. Cigarette smoke can be figured in there somewhere.

Allow yourself to get occasional periods of solitude and fresh air during this season, especially. This practice could help a lot in terms of preventing or reducing symptoms.

Some common notes from allergists for the holiday season include the following:

Holiday decorations may cause problems as they can be dusty or moldy from being stored during the past year. Open the boxes carefully and keep a damp rag handy for dusting purposes. There may be respiratory symptoms if the family sprays with evergreen or pine scent or uses incense or perfumed candles. Foods avoided all year because of allergy should not be taken during the holiday either and be aware that allergenic foods can be disguised.

A well timed visit to the doctor may provide the necessary advice and medication so that the entire season is not spent in bed or in discomfort.

There are a number of reasons why it's tough to make it through the several weeks of the holiday season. It is not only a challenge to our lungs but to our vitality in general. There are a lot of things that can go wrong and it doesn't have to be this way if we can stay in control. What we are looking at is a combination effect of stress, under sleeping, overeating, and yes, over-breathing. The level of outdoor pollen and mold may be low at this time of the year but just about any indoor environment can be a mess.

First, anybody with respiratory problems is just going to have to take it slow as a matter of habit. What this will do is reduce your sensitivity to allergens such as indoor dust, fragrances, cleaners, chemicals, smoke, cooking odors, and any gas or particle that goes into the nose or mouth. Stress, and its attendant negative attitude, enhances a person's sensitivity to these agents many-fold. It will do us all a great deal of good not to take things so seriously.

Now we are relaxed and ready to clean house. Put on some easy music. Unfortuantely, opening the windows with snow and sleet outside won't work, but cracking a few of them just a little for draft purposes just might. Now you can vacuum slowly to reduce the level of airborne dust. Damp dust and organize. Take it slow.

Be aware of the cleaners that you use. Avoid the ones with petroleum distillates and those that have any fragrance, if possible. Baking soda is a great cleaner of walls since it is scent-free and actually absorbs odors.

Or you can clean and deodorize with a mixture of borax and hot water. Polish with a half and half mixture of lemon juice and cooking or olive oil. Use vinegar and water to clean lime from windows and shower doors.

If you have to use chlorine bleach then dilute it one-to ten. It is still effective as a disinfectant on hard surfaces at that concentration.

A lot of people think that if they keep the windows closed and run the heater or air conditioner then this is the same as ventilating a home. This is probably true in the sense that most homes have enough leaks in their ductwork to permit the escape of some bad air and the entry of some outdoor air. This does tend to dilute chemicals or odors within the home.

But for the most part this does not ventilate a home. It just circulates the same air throughout the house again and again. In the winter we can't do much about it. That's why chosing the right cleaning products is so important. Otherwise, the odors from cleaners and polishes builds up and so does the challenge to better breathing. This in turn adds to stress and the cycle continues.

Check the filter to your central air handling system before beginning work to ensure it is new or at least not overloaded.

HOLIDAY ODORS

During the holiday season our vanity tends to run on high. If you have respiratory problems, then you already know that a cosmetology salon is not a place for you. This is especially true this time of the year when they are crowded and the chemical overload in the air is likely to be at an extremely high level.

Cut back on perfumes, colognes, fragrances and aftershaves. This practice will offend less people who come to visit us or who invite us to visit them.

Again, take a little time occasionally to relax and reduce the stress. This is important for your pets as well. Dogs have allergies and respiratory problems as do humans, and getting too excited means a greater number and greater severity of reactions for them as well.

Respiratory symptoms can result if the family sprays or cleans with evergreen or pine scent or uses incense or scented candles. It follows that you should avoid giving items with these smells as presents to those whom you know or suspect have a breathing problem. And a pine

Christmas tree is a joy to millions of persons, but the smell of pine is a cause for an allergic or asthmatic reaction to millions of others. Be aware that this could be a problem for your family.

The amount of time we are exposed to an irritant is called "contact time" or "exposure time." The longer the contact time the worse the exposure. Contact with irritants also includes the time we breathe in smoke from cigarettes, fireplaces and automobile and diesel exhaust.

Exposure time includes inhaling allergens released from our pets. At home it is one-hundred percent of the time. It includes breathing the odors and particles of foods to which we may have an allergy. Studies have now shown that persons with food allergy or asthma can have a clinical reaction just to the smell of the particular food.

Many persons have already turned on the central heaters and noticed the smell. You don't have to live with it. Most of this is dirt and particles that accumulated on the unit itself that have to be burned off. If you haven't started your heater yet open a few windows until the smell goes away.

HOLIDAY STRESS

In terms of challenge for the lungs we need to be more careful now since we have no real control over our environment once we are outside the home. Fortunately the same basic rules apply. This will cut back the amount of time we are hanging around the new clothing or furniture store where formaldehyde and other chemicals are used for permanent press and fireproofing.

Formaldehyde is a sharp smelling irritant. Typical symptoms of exposure to formaldehyde include eye irritation, coughing and headaches. Fresh air usually clears the symptoms in a short period of time.

There are scores of reports in medical journals about perfumes and their ingredients. Just be aware that there is a potential problem here in crowded places and in department stores where free samples are offered.

The air in the central walkways of malls is generally much less hazardous to breathe than that of the individual stores. It's the closest thing to fresh air you'll find for awhile, so take frequent breaks.

If you are buying art supplies or model planes or cars for adults or children be sure to ask for non-toxic glues, markers and other supplies. These are water base products that have been created by manufacturers in recent years in response to consumer pressure. They do not contain xylene, benzene, toluene or acetone. They are available upon request for most of your needs.

Cold weather is an asthma trigger. Some eighty-five percent of asthmatics have allergic asthma. Exposure to allergens, irritants and cold in a hurry-up situation can be a recipe for disaster. It doesn't have to be this way if we can just practice slowing down to be more efficient.

If you avoid certain foods all year because you have food allergy then they should not be eaten during the holidays. Big trouble foods this time of the year are nuts and nut roles, fruit cakes, shell fish and products containing eggs.

It is tempting to try new foods when eating out or at the home of relatives or friends. If you have a problem with certain foods don't be afraid to ask about something you don't recognize. This could save you a lot of grief later.

Chances are you will be spending a lot of time in the car. If you can afford it you might consider getting it tuned up to cut back on exhaust odors, a significant respiratory stressor which can affect the quality of the air within the car as well as in the garage and the home.

CHAPTER 6
PETS

PETS AND YOUR FAMILY'S ALLERGIES

From a number of standpoints, pets contribute to a wide variety of respiratory disorders; most notably respiratory allergy. Antigens from pets include their hair and fur, skin cells or dander, saliva, urine from cat litter boxes, the litter itself, and the indoor tracking of allergens from out-of-doors. Mammalian pets, namely cats and dogs, are among the worst offenders because of their fur, their numbers, and their frequent indoor habitation.

The majority of parents of allergic children believe that the pet is expensive, unnecessary for their child's happiness, harmful to the health, not clean, and a carrier of allergens. In short, they are quite aware of the problems that the pet may create in the household and all feel that their child's allergy is more important than their pet. Given that parents of allergic and asthmatic children understand that their child is reactive to these pets the question then becomes: Why are the pets maintained in the home?

This complex issue has been studied in a number of medical and veterinarian surveys. It was found that there were several reasons for the retention of pets in the home of allergic (asthmatic) children. One reason is tied in with parental resentment toward the child because of the illness. Everyone shares in the inconvenience of seeing the doctor, the medical expenses, the missed schoolwork and changed living patterns- -not being able to do what you want when you want to do it. In many ways this is no different than the resentment felt toward elderly persons living in the home. Very few parents, or children for that matter, felt that removal of the pet would affect the child psychologically.

There were a number of other reasons. Some parents had more confidence in the opinions expressed by their veterinarian as compared with their allergist. This is because many allergists advocate complete removal of the pet from the home, while the vet is more sympathetic. Parents felt that the vet could more easily relate to the pet as a loved one in the family and try to strike compromise solutions with family members so that they would not have to give up this loved one. According to the surveys some parents confide to their veterinarian and pet store owner that the child would overcome the illness eventually, even if the pet was allowed to remain.

Finally, some parents did not like anyone telling them what to do in their home, including their doctor.

Removal of an allergen source and sound medical advice have no substitutes. However, compromise solutions include maintaining the pet in isolated areas of the home or exclusively out-of-doors. Certainly, it should not be permitted in the child's bedroom. Some parents will have their shaggy hair pet shaved down; this could include their cocker spaniel, Persian cat or angora rabbit, according to some pet store owners. Regardless of the route you choose there are still no guarantees.

Take a look at problems associated with some of the more common household pets.

CATS

Depending on where you live fifteen to forty percent of allergic persons are allergic to cats or dogs. Allergy to cats is about twice as common as it is to dogs.

Cat allergen is generated from the oil glands on the skin of cats and can also be found to a lesser amount in their saliva. The allergen is attracted to soft furnishings such as the sofa, armchair, bedding and carpets. It has been found in settled dust and air samples from homes without cats. The antigen is so universal it has been found in every building where it has been sought, including newly built homes, shopping malls, doctors' offices, and even hospitals. Presumably, it is tracked in by persons who have been in contact with the animals.

Cat hair is not allergenic unto itself since we don't inhale the hair. Cat allergen is comprised of two small and potent protein molecules. People who are allergic to cats usually experience a rapid onset of symptoms as soon as they enter a room where a cat is present. Cat antigen is everywhere and takes six months to three years to disappear from a home, even after the cat is gone. It is highest on the bed and mattress. A good thorough house cleaning and scrubbing will remove this highly allergenic protein that sticks to soft furnishings as well as to walls. Studies have found that washing a cat will not remove the antigens from it. An air purifier will not make a difference if you still have the cat since antigen is generated as fast as it is removed. A person who is sensitive to cat antigen can react within seconds to minutes after exposure. No cat breed is less or more allergenic than another.

CAT STATS

. Approximately, six million Americans are allergic to cats.

. Fifty percent of all allergic people are allergic to cats.

. Cats are more likely than any other animal to cause allergy problems.

. 30 percent of all asthmatics are allergic to cats and may have an asthma attack when exposed to cat allergen.

. An estimated thirty-one percent of the homes in the U.S. have at least one cat, for a total cat population of over fifty million!

. One third of those allergic to cats continue to keep them at home.

. Carpets, mattresses and soft furnishings are the major reservoirs of cat allergen in the home.

. Cat allergen is present in a house that has never had a cat.

. Due to the small size of cat allergen particles (2.5 to 10 microns) it remains airborne for long periods of time.

. Persons allergic to cats develop symptoms rapidly upon entering a house with cats in contrast to those who are allergic to dust.

. Dust mite, cat and cockroach allergen are the three main indoor allergens in North America. Exposure to any one of these in early childhood may lead to the development of asthma.

. Significant amounts of cat allergen can linger in a mattress up to five years after the cat has been removed, making impermeable mattress encasings advisable in any house with a cat or where a cat has lived.

. In a house with a cat, aggressive cleaning, good ventilation and removal of soft furnishings can dramatically reduce levels of allergen.

. Washing the cat every month is also useful.

. Children are more allergic to cats than to dogs.

CAT ANTIGEN CLEAN-UP

The highly allergenic proteins are particularly fond of soft furnishings such as sofas and chairs, carpets and bedding materials. It was previously thought that the only way to get rid of them was to wash the home with a three percent tannic acid solution or get rid of the soft furnishings or the cats.

Now, the prestigious medical journal, the American Review of Respiratory Disease, reports that rigorous cleaning of the home coupled with the use of filtration and lots of fresh air is also effective in reducing cat allergen in the home. Some twenty-eight percent of homes in the U.S. have at least one cat. This comes out to fifty million pet cats. Two-thirds of people allergic to cats do not even own one. Studies revealed that homes with cats had from ten to one-hundred times the amount of cat allergen compared with homes that had no cats, and that about 25 percent of the allergen was in particles less than 2.5 microns in size, small enough to enter the deeper lung spaces and cause the rapid onset of symptoms. Larger particles greater than five microns can still induce symptoms after longer periods of time.

Doctors who conducted the study looked at the time it took their patients to react to an experimental dose of cat allergen in the lab and compared it with the amount of allergen found in various homes. They came up with this: it could take as little as twelve minutes to as long as four hours for a sensitive patient to react. They also found that there is much more allergen in rooms where the air is disturbed. This means that, if you have a free standing HEPA filtration air purifier in a given room, the cat allergen will always be there because the clean air that is coming out of the unit will disturb the settled allergen almost as much as it will be removed by the filter.

Most homes have only 0.5 changes per hour or less of fresh air. At ten changes per hour the amount of allergen will be reduced. Even though there would be more turbulence, the allergen would be carried outside.

The study also found that the water filter vacuum cleaner would not be suitable for patients with cats because its use results in a sharp increase in the allergen associated with fine droplets created by the machine.

A recent study suggests that, in addition to genetic factors, exposure to high levels of dust mite, cat or cockroach allergen in early childhood may contribute to the development of asthma. It is well documented that people with immediate hypersensitivity to cat allergen will react within minutes of exposure and at some time develop acute asthmatic symptoms.

A person who suffers an asthmatic reaction when exposed to cat allergen may also experience a delayed reaction four to twelve hours later. This reaction is likely to be much more serious and leave the lungs sensitive to many other non- allergic triggers such as cold air, exercise, and cigarette smoke. These reactions, of course, make a person more sensitive to cat allergen. A vicious cycle. Eventually, if this cycle continues, there may be irreversible damage to the airways.

The preferred method of symptom management is to reduce the patient's exposure to cat allergen through environmental control. If you insist on keeping your cat, try to eliminate soft furnishings and limit the pet's territory within the home. It is advisable to cover your mattresses and pillows with impermeable covers. This is true, in any case, if you have allergic asthma.

Good ventilation can also dramatically reduce the amount of cat allergen in the home. Unfortunately, modern homes that are energy-efficient with a low rate of fresh air exchange trap the smallest particles of cat allergen inside, where they either remain airborne or settle on soft furnishings in a form that can easily become airborne again.

Vacuum cleaners need filters to prevent cat allergen in the carpet from escaping through the exhaust. HEPA vacuum filters are expensive but have been shown to be effective at trapping the small cat allergen particles. A HEPA room air cleaner is able to filter even the smallest particles of airborne cat allergen. It is found to be more useful after the carpet has been vigorously cleaned.

What is the recommended cleaning agent? Published scientific reports claim that a three percent solution of tannic acid will dramatically reduce the amount of cat antigen in the home. This solution is difficult to obtain for the public at large. Fortunately, another published study compared the effects of this solution against your basic thorough house cleaning and found that it was the cleaning itself that did the trick!

Be aware that vacuuming a home with the windows closed will liberate a tremendous amount of cat antigen at face level with all but the best vacuuming systems. And use of a HEPA portable unit will filter the antigen but is ineffective if the cat is to remain for the reason that antigen is generated as fast as it is filtered.

In one recent study, out of forty homes without cats thirty-eight of them had cat allergen in the settled dust. The amount of airborne cat allergen in these homes could also be measured. Although the level of allergen was ten to one-hundred times lower than that present in homes with cats, studies have found that it is common to find patients sensitive to cats who have never lived with a cat. These persons probably became allergic to cats through a variety of methods. Some of these methods may include becoming sensitized though contact with formaldehyde or common outdoor air pollutants.

We can't see cat antigen, we can only keep our home well dusted. This will help. Since any measurable amount of cat antigen might be clinically important, it obvious that total avoidance is not possible. The need for medications should be carefully considered.

In one important study 136 house dust samples were collected from 103 homes across the United States. The investigators were looking for dog allergen (dog hair or dog skin cells). Dog allergen is very specific for dogs and does not cross react with cat, bird, cockroach or mite allergen. Since dog allergen is probably present in half the homes in this country, chances are you have some, even if you don't have a dog! But in homes with dogs, the amount of allergen present is some 360 times the amount compared with homes where no dog is present.

Cat allergen is also present in a very high percentage of homes that have never had a cat present. Cat allergen used to be defined as cat hair until it was thought to be cat skin cells. After that, researchers believed it to be associated with the cat saliva which was spread to the fur by licking. Then it would spread to carpets, furnishings, walls, the air and your lungs.

Researchers have found that cat allergen is actually produced by the skin and stored in the skin. But the hair is also involved. To find this out the researchers removed one-hundred strands of fur and took twenty-four skin samples from eleven different cats. It turns out that the allergen is actually produced by the sebaceous (oil) glands and is stored mainly on the surface of the skin and fur with its greatest concentration at the hair roots. The highly potent and long lived allergen is gradually spread to the tip of the shaft where it may become airborne and produce allergic respiratory symptoms.

This allergen can penetrate bedding and contaminate a mattress where it can remain up to five years. Since cats often remain in contact with bedding for long periods of time the antigen could be a major trigger in families with asthma. For some reason, the antigen remains in the mattress longer than it does in carpets and furnishings. The lesson here is that if you are going to remove your cat from your environment and aggressively clean your home it might be a good idea to get a new mattress as well. And don't forget to clean your ductwork.

One of the most confusing areas of allergy research in recent years has been the debate over the best method to wash cats to remove the highly allergenic cat protein. This potent allergen is associated with the oil glands and skin cells of the animal and with the saliva as well. It is a small protein that is readily airborne and can stick to just about anything, especially soft furnishings. It can last for months or even years, even after removal of the cat from the home.

In the June 1995 edition of the Journal of Allergy and Clinical Immunology (vol. 95, pp. 1164-1171) there is a report that compares the amount of antigen shed into the air by four groups of cats: a control group that had no bathing, those bathed with distilled water and those bathed with recommended products. Six female cats were used for each of the four tests to eliminate sex bias over a period of eight weeks. The cats were washed weekly. This was a highly regulated study with lots of controls and cross checks and was the most thorough study to date regarding the issue of bathing and cats.

Results of the study are that at no time during the eight weeks did any one group show a reduced amount of airborne cat allergen when compared with the control nonwash group.

Many people would rather keep the cat and suffer from allergies. For them having a professional wash and groom their pet has its own rewards. For those whose affliction is serious enough, removal of the cat from the home, together with reduction in soft furnishings and damp dusting is still the best method of indoor cat allergen reduction.

DOGS

Skin cells of dogs are allergenic and long-hair dogs have been reported to be less allergenic than short-hair breeds. This is because they shed dander into the air. This is relatively unimportant, other than in a scientific sense; a person who is sensitive to dogs will probably react to any breed. From the human standoint it is important to have your pet groomed and well cared for to maximize its health and minimize your own exposure to allergenic animal dander (skin cells) and hair. Regular care keeps the coat healthy and reduces the shedding. This may not make any difference as far as cats are concerned, since it is not the dander that is allergenic.

If you are looking for a breed that is less allergenic to humans, try a poodle or a terrier. These dogs shed less hair and dander than long hair breeds.

Short hair breeds shed more dander than long hair breeds. And it is the dander of dogs that is allergenic. Hair is long anyway and it's tough for even one hair to get into the respiratory tract to cause an allergic reaction. If we analyze the air of homes where long hair dogs are present we do not find short pieces of hair that might have broken off and capable of entering the sinuses or lungs.

BIRDS

Birds produce a powder on their feathers that is inhaled, yet it is their droppings that are the main troublemaker.. Fortunately, birds are contained in a cage in a particular room, while cats and dogs are free to contaminate the entire home. Birds also shed tiny feathers that become airborne, but are not inhaled to the extent that they become a problem. Pigeons on the roofs of buildings can roost around fresh air intakes and have been found to be a problem because fragments of their droppings go into the duct work.

Like cat antigen, bird antigen will remain in the home long afte the pet is gone, and is found primarily in the soft furnishings and deep pile carpets. Normal cleaning of carpets does not remove the antigen. Clean the cage once weekly.

There are a number of diseases that both wild and cultivated birds transmit to humans. The birds include chickens, turkeys, starlings and others. The diseases go by the exotic names of cryptococcosis, histoplasmosis, Newcastle disease, cryptosporidiosis, salmonella and psittacosis. The latter two are transmitted by bird droppings.

There are also millions of pet birds in the United States. Most of them are parakeets, canaries and finches with lesser numbers of cockatiels, and parrots.

Domestic birds release antigens into the air. These are responsible for a variety of respiratory ailments. Health problems can persist in patients for months even after the bird has been removed from the home. The most common of these respiratory ailments is allergic rhinitis, asthma and hypersensitivity pneumonitis.

There are two main sources of antigens from birds. The first is their feces, fragments of which become airborne. The second is powder on their wings that also readily enters the airspace. Minute birds' feathers that are released are probably not as important a source of respiratory problems as the other two. This is because of the potency of the antigens and their microscopic size.

Numerous studies have found that various health problems exist long after the bird is removed from the home. Because of this it was thought that the antigen(s) was long lived, similar to cat antigen.

The difference is that cat antigen is found throughout the home at a high level. Bird antigen, however, is found predominantly in the area where the bird is kept and in lesser amounts distant from the pet. This is reasonable since the bird stays in one location. The exception to this is when the bird spends a great deal of time away from the cage in other areas of the house.

Antigens associated with the bird can remain in the home for as long as eighteen months or more after the bird is removed. This means that clinical symptoms can persist during that time period as well.

If you have a sensitivity to birds here are some tips that you should consider:

. If possible keep your bird in a room that has wooden or tile floors. This will help you keep the room clean (two - three times weekly). It will also prevent the spread of antigen elsewhere in the home.

. Long-pile carpet within the home will store the most antigen.

. Clean the cage at least once a week.

. It is unfortunate, but studies have shown that normal cleaning of carpets does not help remove the antigen. However, deep steam cleaning may help because it can destroy the protein antigen.

. Wild bird excrement deposited outside the home may be a source of antigen in the home if it is tracked indoors.

. Do not kiss the bird.

HORSES

The dander of long-hair horses is just as allergenic to humans as the dander from short-hair horses. But since long-hair horses shed less dander into the air the reaction to them is less. This helps explain why many persons report that they are not allergic to long-hair horses. This is similar to the situation with dogs.

Tens of millions of Americans own, ride or come in contact with horses on a regular basis. In a situation similar to that of dogs scientists have known for a number of years that the hair and dander of horses are allergenic. (Dander is also known as skin cells.) Are some breeds of horses more allergenic than others?

Soon it will be one-hundred years since modern science began its study of allergy and dander. We still have large holes in our knowledge.

Some people are highly allergic to all horses. But for years, people have been telling each other and telling their doctors that they are allergic to some horse breeds and not to others. The investigation into this matter was a case of many incomplete studies laying the groundwork for a major study.

In some recent major research, a number of horse breeds were compared to see if the dander from one breed was more allergenic that that of another breed. This was a Swedish study, and Swedish researchers are among the best in the world when it comes to allergies—whether it be homes, schools, people, animals or test tubes. The horse breeds that were studied included Swedish Halfbreed, Haflinger, Arabian thoroughbred, Bashkir Curly, Russian Bashkir and American Trotting Horse.

Remember that an antigen is a substance than can cause an allergic reaction. In this study no antigens were found that were specific to certain breeds of horses. In other words if you are allergic to one breed of horse you are allergic to them all. All the horses in this study carry similar antigens.

The dander of long-hair horses such as Bashkir is just as allergenic as the dander from short-hair horses, such as Arabians and thoroughbreds. But since long-hair horses shed less dander into the air the reaction to them is less. This helps explain why many persons report that they are not allergic to long-hair horses.

PETS HAVE ALLERGIES, TOO

Dogs have the same allergies as do humans (for example pollen and mold), but they manifest their reactions by scratching, while we sneeze.

Allergies in dogs are more severe when they begin at an earlier age. When sensitivities develop they can be to pollen, mold, house dust, food, chemicals, insect bites or other factors. As with humans, dogs inherit the ability to be allergic.

Unlike humans, respiratory allergy in dogs is commonly manifest by severe itching and scratching of the skin as well as by licking and biting of the paws. Other symptoms may include sneezing, coughing, tearing or diarrhea. If your dog starts to scratch about the same time each year he may be suffering from a skin allergy due to breathing pollen (February, March and April for trees, late spring and fall for grasses, August and September for ragweed and other weed pollens). A high percentage of dogs will develop allergic clinical symptoms at one time or another.

Reactions to mold can be seasonal, as well. Allergic reactions to mold also occur indoors in the damp basement or if a carpet is contaminated. In this case, see if the pet's initial reactions occur in a particular room. Also watch for licking at a runny nose, rubbing an itchy face on the carpet and other symptoms mentioned above as clues to an allergy problem.

Respiratory allergy in dogs can be confused with flea and tick allergy and irritations to the eyes and ears; it is most common in Dalmatians, Poodles, Cocker Spaniels, Golden Retrievers and Wirehaired Fox Terriers, but all breeds can be affected.

Food allergies in dogs and cats is not uncommon, especially with today's processed and preserved foods. Food allergies tend to occur year-round rather than seasonally as with pollen and mold. Severe itching is very common here and one must rule out allergy to other coat irritants.

Many persons will use their own shampoo on their pet. This practice may tend to dry the animal's skin since an animal's shampoo has to be balanced for the pet's skin.

What can the pet owner do to determine if the pet has an allergy?

According to veterinarians, the first step is to educate yourself as to the signs and symptoms of pet allergies compared with other animal disorders. Your local veterinarian can provide you with literature in this regard. For example, dogs will vigorously scratch their skin in response to a respiratory allergy, unlike humans.

Then do some detective work. Perhaps lice or ticks are present in the coat that is causing the itching. Is there a rash around the snout due to sensitivity to a new food dish? Are there new plants in bloom or is do you have a home with a lawn? Similar to humans, pets can react to the chemicals in a new carpet; they may have difficulty breathing or their eyes may tear. This is also true when they are exposed to cigarette smoke.

If it is determined that an allergy is present in your dog, for example, and it is manifest in the skin, a bath will be helpful. Use an anti-allergy shampoo and bathe the animal in cool water. Warm water will irritate the skin condition. Leave the skin in contact with the shampoo for ten to fifteen minutes. This will reduce the inflammation and help to rehydrate the skin. After the shampoo is rinsed follow the treatment with a moisturizer (no fragrances please). For treatment of dogs tar-sulfur shampoos are available with prescription. Be careful. These can be fatal to cats since cats are more sensitive than dogs to most medications and treatments.

Dietary supplements will help. Use Omega 3 and Omega 6 fatty acids; the liquid works better than the powder. This will affect the sheen of the coat from the inside out. These are anti-inflammatory fatty acids and will serve to reduce the itching and scratching. This means that your own exposure to the allergenic dander and hair will also be minimized. This is an important consideration.

Most veterinarians will perform allergy testing on your pet, either skin tests or blood tests. Once the sensitivity results are known pet owners may be able to provide desensitization shots to their pets at home. Check with your vet.

As in the case of human allergies your vet may call for more aggressive treatment of the pet's problem, such as by the injection of corticosteroids. Also, ask your vet about ticks and fleas. Treatment by a grooming shop may be inexpensive, but may cause problems to the pet.

CHAPTER 7
MACHINES AND DEVICES IN THE HOME

HUMIDIFIER

If you wake up with a dry scratchy throat or your skin and scalp itch, then you may need a humidifier. Turning on the furnace in the winter dries out the indoor air and may lead to one or more of these problems.

A humidifier can help if you remember that humidifiers can have their own problems. On the other hand, if you have asthma, it may be just what the doctor ordered.

There are a number of things to consider when shopping for a humidifier to avoid getting into trouble. Since they become contaminated with mold and bacteria easily, they must be easy to clean. Sometimes the manufacturer does not give the best recommendations regarding what to use to clean the device. You could leave a toxic residue that becomes an aerosol that is breathed. Stick with something simple such as vinegar or a half lemon.

If you buy an ultrasonic humidifier, beware of the following problems: Ultrasound causes the formation of micro-water droplets. If contamination is present, then the bacteria and mold also are sonicated into micro fragments that are inhaled more readily. Hard water and residual lime also can be sonicated, and symptoms of lime inhalation can be similar to those of microbial inhalation. So keep the unit clean and use distilled water. This can get expensive.

A humidistat on the machine is a device that tells the relative humidity, but only near the machine, where it is attached. You are the best humidstat. Do not put too much faith in this gadget.

The wick type of humidifier is safest from the standpoint of microbial growth, although somewhat noisy. Impeller types have fan blades that sling water into the air. They are easy to clean but when contaminated can become a real problem. They require soft or distilled water and this can cost more than the machine itself over a year's time.

Steam-mist types use a lot of energy. The heat will kill microorganisms so filters aren't needed. However they can still emit dead microbes when contaminated. There's not much difference from an allergy standpoint.

Table top humidifiers should have a portable tank with handles or side-grips for convenience and a large enough capacity to humidify a medium-to large-size room. Console models should have casters or wheels and a switch that shuts off the unit when the water tank is low or empty. This size unit is touted to humidify an entire home and will often have a humidistat. In any case, buy a unit that opens wide for ease of cleaning (about once weekly).

Be conservative in your efforts to add moisture to your lives. Remember that dust mites are prolific when the relative humidity gets above sixty-five percent.

PORTABLE AIR PURIFIER

Most will provide a pre-filter for larger dust particles, a carbon filter for odors, and a high efficiency HEPA filter for fine particles. Many of these units will run the carbon first as a prefiliter. This is wrong. The reason is that the charged carbon particles will be blocked by dust in the air, the ability of the carbon to remove odors from the air will end very quickly.

You can use carbon as a pre-filter to protect the main HEPA, but if you do, just don't believe too much in the odor absorbing ability of the carbon. One reason is that the dirt will attach to the carbon particles and keep them from absorbing odors. Another reason is that the amount of carbon used in these pre-filter sheets is generally so small that their life is extremely short. They can be used up after absorbing the smoke from a single cigarette. Thus the carbon has to be changed much more often than the manufacturer recommends.

A word of caution: Before you fall in love with one or more filtration type products, be sure you need one. Most of these products will do what they claim to do; at least for a limited time. The question is, will they reduce the allergen amount enough to result in a reduction of your symptoms?

There are dozens, if not scores, of companies making these units. But there has never been a single published study in a medical journal that has shown a lessening of symptoms in the home when the units were used. And this is probably true of vacuum cleaners as well!

Suffice it to say they are only part of the solution, not the answer to the problem. You must be the other part and maintain a clean dust-free home.

There is no doubt that the air purifier industry is huge. Like everything else, there is a right way and a wrong way to do things.

The ideal air purifier should have several qualities. Solid construction is one of them. Many people prefer a unit constructed of metal, rather than plastic. This tends to make them more duarable and longer-lasting. (It also tends to make them heavier. This is not a plus for many people.) Another feature is the sound factor. The purifier should have several settings for speed and be relatively quiet when in operation. The consumer will soon get used to the sound of the motor, however, and this will tend to block out background noise.

The ideal air purifier should have lots of carbon. So much the better if it also has additives such as zeolite and potassium permanginate. Zeolite is crushed volcanic rock that has a tremendous ability to absorb formaldehyde, hydrogen sulfide (from sewer gases) and ammonia smells. Then the carbon takes care of the rest of the smells like glues, perfumes, paints and whatever else causes odors in the home or business.

The unit should have a pre-filter with lots of square feet of surface area and a large HEPA filter to take out any particle down to almost the smallest size.

A long lasting warranty should come in more than one size and have a long-lasting warranty and a return policy. In addition, the large size should have wheels..

It is important for the consumer to take notes while shopping for an air purifier. Claims are easy to make, true workability is a different matter. Compare the efficiency of one brand with another, but don't expect to filter the entire home with a portable air purifier.

VACUUM CLEANER

So you think you need a new vacuum cleaner but you don't know what to buy. Before you get too involved in the shopping process take another look at your old unit. If it is a canister and the only problem is that it leaks dust into the air, try using duct tape to wrap the area where the hose fits into the canister. If it is an upright then you might want to call around for better quality bags. If you have made up your mind to get a new vacuum cleaner, never mind the 21st Century, first review some basics.

Be wary of too many gadgets. You need something that is versatile, easy to roll around and has a good strong suction. And price is not always a good way to compare different models.

A current Consumer Reports Buying Guide suggests that the middle range of prices will provide a good vacuum cleaner for the average person and the average need. At that price one can get a unit with a cleaning path of some fourteen inches, good quality bags and a wider range of tools than less expensive models. The same guide notes that the popular Oreck vacuum rates high among vacuum cleaners for its price range.

Uprights generally work better on carpets, but many also perform well on bare flooring. If you need a lot of attachments for specialized cleaning jobs such as underneath beds and furniture and have a lot of bare flooring, then a canister will probably work best for you. Canisters are also quieter than uprights and do a better job on carpeted stairs.

Cord length is a consideration. The average is twenty to thirty feet and many canisters have retractable cords, although uprights are easier to store.

If you opt for a unit with no bags, remember that you will still have to empty the bag. This can be a problem for some people.

Unless you decide to buy a vacuum cleaner with a "full-dirt" indicator, you will have to change your bag on a regular basis. This latter is not preferred, as it is just one more thing that that can go wrong. Better than the user be in control and remember that when a bag that gets too full, the suction power is reduced.

HEPA filtration sounds good, but there is no way to tell if the unit is constructed good enough to tell if the filter is doing what it is supposed to do.

Be aware that if you have small children in the home there can be small objects on the floor. This is important because the blower fan in most vacuums is made of plastic, and it is this fan which drives the dirt into the bag. It is also breakable. Some units are constructed such that the dirt is filtered through the bag before reaching the fan. Hence, no breakage.

A suction control is present on canisters, used for doing draperies and specialized objects. Uprights frequently allow for increased suction by lowering or raising the unit from the carpet or floor surface. This is fine, but try to chose a model with large rear wheels to make the pushing easier.

Here's an important point. Suction power, that is, the ability to suck up dirt, is only one factor. A badly designed unit can off-set suction power and many lesser expensive units can clean just as well as the most expensive models out there.

Another important point: Successful cleaning is shared between the cleaner and the "cleanee." The carpeting industry tells us that slow cross-wise vacuuming is a lot more important than fast back and forth motion. Why? Because the first removes dirt most effectively, and the second scatters it into the air to be inhaled.

Make sure you purchase an attachment that can get those areas between the carpeting and the walls and between the sofa pillows. You will need some sort of hand attachment to get the surface of the sofa and arm chairs where an incredible amount of dust accumulates in a short period of time.

In general, canisters are better than uprights, but virtually all name brands of canisters rate the same, and virtually all name brands of uprights rate the same. So decide how much you want to spend and what features you need according to the needs of your household. Make sure you can purchase quality bags for the unit and find out what the cost is for those bags, since that will be an occasional expense.

REFRIGERATOR

If you want to lessen the load on your lungs, then clean the refrigerator. There are ten or more areas of the unit that can catch you by surprise, and need to be checked on a regular basis. These include handles and outside surfaces, the refrigerator top, the door seals and paint, the vegetable and meat drawers, the food itself, the drain hole, the drip pan, the floor beneath the unit, the fan, and wall behind it. If you have a family, odds are that you've got some work ahead of you.

As everyone knows the top of the fridge accumulates a lot of dust and grease because it is above eye level and is frequently located near the stove. This area should be washed with a damp warm wash cloth or sponge and dishwashing detergent. Then wash out the sponge and put it in the microwave on high for sixty seconds to sterilize it. The sponge should be microwaved daily, in any case. Don't neglect to clean the disease-laden handles and front and sides of the unit.

The rubber seals around the doors of the refrigerator frequently have food particles and mold growth. Cladosporium (Hormodendrum) is the common mold here. This allergenic mold will grow on the door seal, and spores will be liberated into the airspace when the door is opened. Use a damp soapy washcloth or cotton swab to clean the grooves. The mold will come off readily. Don't forget to check the seal around the freezer door as well. Finally, mold growth can occur on the paint where the doors meet the body of the unit.

Pull out the vegetable and meat drawers. Take out the food and clean out any pools of water within and beneath the drawers. This water is usually contaminated by microbes that have adapted to colder temperatures. Examine the food itself and discard soggy and decayed vegetables, fruits, breads and other suspect food items.

If you are thinking about eating the moldy cheese to get free antibiotics, forget it. Although the mold that contaminates cheese (and bread) is usually Penicillium, you will breathe in massive quantities of spores in the process. These spores will be liberated into the kitchen and home. In addition, you may get a dose of fungal toxin instead of penicillin antibiotic. Penicillium mold produces both antibiotics and toxins. It could be the equivalent of eating poisonous mushrooms.

Open old jam jars and check for mold growth on the surface of the product and where the lid connects to the jar. If the jam has an alcohol smell discard it. Yeasts are at work here. If the pancake syrup has something growing on the top of it, discard it. In fact, when food gets low it might be a good idea to empty the entire refrigerator and clean it completely with a solution of baking soda; one-half cup in a gallon of warm water. When you put the food back in you may want to include a small dish of baking soda to absorb odors.

Make sure the hole to the drain pan is open and unclogged. This is a hole for runoff of water condensation. This water goes to the drain pan underneath the refrigerator. In older units this hole is located inside and at the back of the unit between the vegetable drawers. A pool of standing water in that spot is a sure indication that the hole is plugged. Bacteria will be growing in this water.

Soak up the excess water with paper towels and discard. Usually the hole can be opened with a wire.

Pull out the grate at the bottom of the unit and take a look underneath. Most sub-fridge areas have dust balls as well as fine dirt, pollen, spores and allergenic insects and their parts. You may find some surprises as well. Warm air from the motor causes the finer particles to become airborne thus exposing family members to them. That's a lot of exposure considering the number of family hours that are spent in and near the kitchen. The unit will have to be unplugged and pulled out and the floor carefully swept and then damp-mopped.

While the unit is out be sure to carefully clean the area around the fan. This cleaning will help avoid expensive repair bills since the fan will cease operation once it becomes clogged with dirt. Once the fan stops so does the refrigeration.

Once the refrigerator is pulled out, don't forget to wipe down the walls beside and behind it. And be careful not to injure the cooling coils while you pull out and return the unit. Gently clean the coils while you are at it.

Next, while the bottom grate is off, pull out and wash the drain pan that is present in most older refrigerators. You will need to pull out the grate at the bottom of the unit and, unfortunately, get down on your hands and knees to locate the drain pan.

CENTRAL AIR CONDITIONING SYSTEM

Many people believe that a better filter will take out bad smells from the home. This is mostly not true. Gases that are attached to dust-like particles will be filtered. The rest of the gas molecules are mixed in with the air. These are the main part of the odor and will pass through even the best paper or fiber filters.

Either remove the source, or go to carbon or zeolite filters to remove odors, if that is your concern. In any case, all homes need to be completely aired on a regular basis, once or twice monthly, depending on your indoor problem. Sometimes it might be after several months depending on the local climate.

Air conditioning units in commercial buildings have a certain percent of fresh air that is pulled into the building. This air is mixed with the inside air, cooled and filtered. Sometimes even this is not enough to dilute smells such as from the adhesives used to glue down new commercial carpeting. In the home the AC unit has one-hundred percent recirculated air with no fresh air other than what you let in. Odors are mixed and spread throughout the home to be adsorbed into soft furnishings, wallboard, wallpaper and clothes hanging in the closet.

When air is recirculated in an AC home, it travels through the duct system. Since most home duct systems have leaks there will be some air exchange with the outside. This is unwanted. This means that outdoor pollutants will be pulled into the home. A lot of the particles will be filtered, but there are several problems with this air mixing in the home. First, it is not energy efficient. Also, real contaminants such as mold or automotive gases can be pulled into the home to be breathed before they are filtered.

Apartments are notorious for permitting gases such as cigarette smoke or cooking odors to travel from one unit to another. There are a lot of ways this can happen. One of the ways to reduce this problem, whether you are the giver or receiver, is buy foam inserts to fit behind wall sockets. At your local hardware store you can purchase plugs for the outlets. Put them into unused outlets.

Then get some aluminum tape and seal the areas under the sinks where pipes go into the walls. Look for other areas where smoke can leave or enter your apartment. It all helps.

WINDOW AIR CONDITIONER

The most common molds blown into a home from the window AC are members of the Penicillium and Aspergillus groups.

In fact, the common green bread and cheese mold can grow as fast in the refrigerator as outside of it. These are common bad guys of the molds. They grow at all temperatures and a wide a range of humidity. Their spores are extremely small and extremely numerous. So they can enter the deeper lungs to cause problems with the allergic, asthmatic and immuno-compromised patients. It has been documented many times over that this last group of people can actually have Aspergillus growing inside of them when they are exposed to this group of molds. These molds love a dirty filter because they love dirt. Keep that filter clean.

The evaporator coil is inside the unit and is the cold part of the AC. The AC removes moisture from the air, and during the rainy season, the air is very humid so a lot more moisture will be removed. This moisture forms on the coil and can harbor mold growth.

After water condenses on the evaporator coil it drips into the drain pan. The pan has a hole in it so the water can drain out. Sometimes debris can clog the hole and water will accumulate in the pan, similar to the interior of an older refrigerator. Microbial growth will occur in this mess and so will its smell.

The window unit should be level, otherwise moisture could run into the home. This is another problem that will occur if the drain pan is clogged.

EVAPORATIVE COOLERS

Tens of millions of people in the southwestern United States and in desert areas of the pacific northwest use the evaporative cooler (EC), in homes, businesses and schools. There are just as many people who suffer from allergies who believe that air conditioning (AC) is better for their health and they would use it if they could.

In truth, the jury is still out on this issue. The reason is that in most cases of indoor allergy where the EC is involved it is used improperly. The EC is so widely abused that it is no wonder that more people aren't sick from it. Maybe you are ill from its use but don't know that the cooler caused your problems.

As we all know the EC can load up with lime, bacteria, mold, leaves, and yes, mosquitoes, when it isn't cleaned on a regular basis. There are reports of schools that have turned on the cooler for the first time of the year and have had live mosquitoes sent into the classrooms at high speed to panic teachers and students.

It is a safe bet that many of the same people who complain about indoor illness from the EC haven't cleaned it in a long time.

When properly used, the EC will blow particles of dust, dirt and accumulated allergens out the windows and provide good cooling at the same time. Proper use of the EC calls for getting the pads very wet before turning on the blower. This keeps hot dry flakes of lime from blowing into the home when you first turn it on. When inhaled lime dust can mimic allergies and may precipitate an asthma attack. Proper use of the EC also calls for paying attention to your duct vents and how much the windows are open.

Close or decrease the vents in the room(s) you are not using. This will give more force to the other vents. You may be able to use a lower power setting as a result and save money. Redirect any register that blows air right on your head. This factor alone can tend to aggravate allergies and asthma due to excessive cooling. Even otherwise normal folks can get what is called a "fan hangover" if this is allowed to occur. If you have a fan pointed at your head while you are sleeping you can wake up feeling very groggy, stuffy and wiped out. Same thing with the cooler vent.

Also, avoid pointing the register at walls or furnishings since moisture can tend to accumulate on these objects under the right conditions. Pictures, walls, even the side of refrigerators can become contaminated with the growth of mold. The spores are released into your breathing space.

There are a lot of factors that affect how open a window should be for the best cooling effect with an EC. One rule of thumb is to open a window just past the point of whistling. Another rule of thumb is to open the a window from four to six inches at either end of the home. Experts disagree about this but they do agree that doors within the home should be open. Each person must experiment since house and cooler sizes are different.

If a window is open too much there will not be enough positive pressure in the home and particles from the outside will enter the home. If a window is not open at all then there will be very little cooling and allergenic particles in the home will build up.

The evaporative cooler has two big problems: microbes and lime. The water reservoir in the cooler usually contains a great deal of dirt, leaves, bugs and nutrients to support a large population of microbial life. Just in terms of bacteria, the water can harbor millions per milliliter, enough to sour milk instantly if they were the right kind.

Some of the cooler-water bacteria are potentially harmful and are certainly capable of causing some respiratory irritation and infections, especially in the elderly and more debilitated. Infections from coolers are more likely to occur when we are run-down or have some underlying infection. Just two of these disease associated with evaporative coolers are humidifier fever and hypersensitivity pneumonitis. You don't have to be allergic to get these diseases.

Mold becomes an important problem as well. Two of the most dangerous mold types to commonly inhabit the cooler are Aspergillus and Fusarium. These allergenic molds are transmitted into the home.

The color of the water in the cooler reservoir is not a clue as to its relative cleanliness from microbes.

The accumulation of lime prevents the pads from absorbing water and evaporating it as water vapor. This cancels out the cooling effect and also allow the entry of free water droplets into the home. If an indoor register is pointed at the wall or at furniture or the side of the refrigerator, then this is where the free water will accumulate and this is where indoor mold will begin to grow.

The cooler needs to be serviced about every eight to ten weeks. Scrub out the lime, change the pads and service the belts if necessary. It should be serviced more frequently if there is a lot of blowing dust to contaminate it. After this period of time microbial growth becomes significant. Just servicing the cooler at the start of the summer and at no other times means that you will be at risk and also have some lousy cooling as time goes on.

Mark R. Sneller, Ph.D.

CHAPTER 8
HOME CONSTRUCTION

AIR LEAKS

This segment refers to air conditioned homes throughout the year, whenever the system is on, in evaporative cooled homes, and in any home when ducting is involved.

There are several reasons why we need to know the amount of air leakage in our homes. We need to know where allergens and dirt are leaking in, we need to know for the sake of energy efficiency, and we need to know if we are getting enough fresh air to dilute any pollutants and odors in the home. Odors can come from a wide variety of sources.

There are a number of areas where air can enter or leave a home when the home is closed. This is especially true if you have leaky ducts. Doors and windows are obvious areas, but we want "some" leakage here because we need the fresh air exchange. This is compared to a door to the garage that is not tightly fitting and may permit polluted air to enter the home.

If you see spider webs in the corners of your windows this is a sign that air leakage is occurring here. It is the area where the spider sets up a home because other spider-edible creatures will enter through the opening.

Other areas of leakage include the chimney when the damper is left open.

Plumbing fixtures have rings set against the walls. If the rings are not set tightly then air from the inner wall spaces can enter the home and vice versa. This includes plumbing under the sink as well as shower pipes that go into the wall.

It is common to find poorly sealed ductwork fittings off of the main duct trunk. You can determine this by removing a register and seeing if there is a tight fit of duct against the wall. Poorly fitting registers is common to new homes. In older homes the tape that holds the ducting to the air conditioner and heater can become worn and will permit the entry of outside, attic, crawl space or garage air into the home.

Recessed lights and electrical fixtures frequently leak debris from inner wall spaces, attics and crawl spaces into the home.

Check the inside of your medicine cabinet. If it has metal brackets for the glass shelves, then these will open into the walls. It is a source of minor leakage into the inner walls of the home.

Recessed sliding wooden doors have huge leaks into the inner wall spaces. There is nothing that can be done to fix them.

If mold contamination exists within the framing of a wooden home it is not hard to see the many ways that spores, or at least, musty odors can enter the home proper. When there is positive pressure within the home this is not a problem. When there is negative pressure then odors will enter the reduced air pressure that is created, similar to outdoor breezes that move into an area of low pressure.

Air pressure changes are common within homes. If there is a leak in the supply side of the system, that is, from the air handler to the interior space, this can be a serious problem. The result is negative pressure inside the home and the following pollutants can be drawn inside: radon gas, formaldehyde, pesticides, car exhaust from the garage,, household cleaners, mold and volatile organic compounds from several sources.

Negative pressure can occur under a number of other circumstances, as well. For example, suppose you maintain the bedroom as a clean room and the door is closed. If there is no air supply register in the room or the register is closed then negative pressure in the room can occur. This causes air to be pulled in from other rooms and the room is no longer a clean room.

You can frequently tell which way air is flowing in a room by looking at the bottom of the door. You may see a strip of dirt along one side of the door. Or you may see this strip of dirt in the carpet beneath the door as air passes under the door. More dirt is present on the entry side.

Usually the worst area for air system leakage is the return air duct. Remove the return air register and take a good look behind it. You will probably see dirt, holes in the wall and construction material. These all affect your breathing.

The closet that houses the water heater should be treated as an outside closet and should be tightly sealed. Roof vents serve to relieve excess gas that is generated, if the unit is gas powered. Indoors, loosely fitting doors means conditioned air will pass through the opening and out of the roof vent. This is not energy efficient.

A home can be checked by an Infiltrometer, a computerized device for determining where the leakages occur within a home and the extent of the leakages. Under normal circumstances, a 1500 square foot home should have about 1.2 square feet of total leakage area. On average, this leakage would exchange all the air in such a home with outside air the recommended eight to ten times each day.

Too much leakage is energy inefficient, causing too much dirt to enter the home. Leakage occurs in duct systems and from ill-fitting doors and windows. The latter can be fixed with a few dollars worth of weather stripping obtained from the hardware store. Some of the duct leakages may be fixed by removing the registers and sealing any open space that exists between the duct and the wall.

NON-TOXIC BUILDING MATERIALS

There is an ongoing interest in the relative toxicity of construction materials. Let's review a few of the most common construction and repair materials and their irritant or non-toxic nature.

CONCRETE FOUNDATIONS: This is usually acceptable after curing unless it is treated with formaldehyde, petroleum oils and detergents. Slabs may contain pesticides that out-gas into the carpeting and the home when cracks develop. Check with your builder regarding details of the materials he is using. He may buy in bulk and may not know the answer. In this case the supplier must be contacted.

MASONRY FOUNDATIONS: These are more permeable than concrete to moisture and radon gas.

STEEL FRAMES: These may be coated with oily residue from factory. The residue can be removed with detergent.

GLUES: Normally used are white glue and yellow glue containing acrylics, casein, and/or vinyl acetate, but no petroleum solvents. Wallpaper glue usually has a starch base and is okay; but some commercial glues contain mold retardants and pesticides. Incidentally, old wallpaper, on analysis, has been found to yield pesticides such as Diazinon®, Dursban®, Heptachlor®, Chlordane®, and Mirex®. These have been absorbed from the indoor air.

Wall tile adhesive: unacceptable during application since it contains toluene, benzene and naphtha. Once covered and sealed by tile and grout its presence is not a problem. Use plenty of fresh air when applying it.

Epoxy glues: These can be problematic in terms of long-term out-gassing.

Area rugs: The safest are cotton and wool with untreated natural fibers; some may contain chemical dyes, residual pesticide and mothproofing. The worst floor coverings are glued-down carpeting, soft vinyl tile, self adhesive tile, simulated wood flooring and foam rubber carpet pads.

Cabinets: The best have a metal frame and doors; the worst are made of hardwood veneered "cabinet stock", standard particle board and vinyl "imitation wood" panels. These latter out-gas gases from plastics and formaldehyde.

Small appliances: Most give off ozone, oily fumes or create secondary odors; these include, blenders, refrigerators, electric dryers (inside vented) and irons.

Lamps: Metal and glass lamps are best. Fluorescent lamps often contain hard plastics which out-gas somewhat, otherwise they are fine. Avoid lamp shades with plastic liners. Bare light bulbs are hot enough to burn microscopic particles such as dust, plant and insects parts.

Pipes: Copper pipes with mechanical joints are better than copper pipes with lead soldered joints; galvanized steel is lined with zinc which can enter the water supply upon reaction with chlorine; PVC contains vinyl chloride, a human carcinogen; polybutylene tubing will spring leaks, which leads to water damage and subsequent mold growth.

Silicone sealants: ,The clear variety is best once it dries. Linseed oil putty is also widely acceptable. Silicone tub caulk, however, contains mildew retardants, which may be a problem for some people. Butyl rubber and acrylic sealants are the most unacceptable.

FLOOR COVERINGS

There are a lot of choices of floor coverings and different people have different needs. What we all have to be concerned with is this: Which ones are the easiest to maintain for people who care about their respiratory health. Look at the pros and cons of these choices.

Carpets include what you can see but also include the hidden backing and pad. The term "carpeting" refers to all three. Carpet is easiest to clean when it is of the commercial variety and low pile. When wet, it will be free from mold growth if it is made of nylon, rayon or other synthetics. The pad can grow mold if wet since it is frequently a cellulose base product; is a good food source for mold.

Synthetic carpets are stain resistant. The shallower the weave the easier it is to vacuum and will hold less dirt, pesticides and other contaminants. Some problems here are that you will have to search for a non-toxic glue to hold the padding to the slab. Also, some people react to nylon by contact.

After installation new carpets should be vacuumed very slowly to pull up loose fibers that can otherwise be airborne. Once airborne the fibers will be inhaled or burned off in the heater to create toxic gases and particles.

Astroturf is made of polypropylene, usually with a vinyl or rubber backing. It is relatively safe but obviously it has limited uses indoors. Indoor/Outdoor is similar. It can be used in playrooms or in the patio; it shows dirt readily. It is fairly easy to clean but does not wear evenly.

Vinyl tile will go over wood or concrete nicely. There are advantages here. Damaged tiles can be replaced. An endless variety of patterns are available. You can install it yourself. It keeps down the humidity in the home by acting as a vapor barrier on top of the concrete slab. This is more important than might seem at first glance. Cracks in the slab will tend to allow under-home pesticides to migrate upward into the living area under certain conditions. Vinyl tile prevents this from happening.

If you decide to use vinyl tile you will still need to deal with the glue. Since the tile will be laid over the glue the off-gassing from the glue will be minimal. Fresh air will help here.

Keep the patterns simple. The more complicated the design, the more grooves will be in the tile and the more dirt will be retained. Also, it will still have to be waxed occasionally. After installation, mop with warm soapy water to remove excess new product chemicals. These are volatile; that is, they can be airborne readily. Make sure you have several squares and some glue left over for replacement purposes later.

The basic idea is that you want a flooring that is very easily applied, retains and shows as little dirt as possible and can off-gas very quickly during and after installation. Dirt includes adsorbed pesticides, as well as plant parts, pollen, mold, food particles, animal dander and outdoor allergens that are tracked into the home.

Concrete should be sealed. There are even artists who can paint marvelous scenes on your concrete flooring. But concrete can be cold, and it cracks. Cracking can allow under-the-home toxics to come indoors. It will also have to be sealed for ease of cleaning.

Ceramic tile costs about $2 per square foot and up to install. It looks good, can be damp-mopped and is easily swept. Broken pieces can be replaced and can be used in combination with other types of floor covering such as low pile Berber carpeting. Ceramic tile has is a great non-toxic, low dust floor covering.

Saltillo tile is similar to ceramic tile but requires waxing and fairly regular care. As such the level of volatile chemical irritants in the home could be elevated from the care products.

Linoleum is the grandfather of vinyl. It is made of cork, wood and mineral chips bound by hardened linseed oil on a canvas backing. It is brittle over time and is less stain resistant and chemical resistant than vinyl. Both can be cleaned with half a cup of white vinegar in a gallon of warm water.

Area carpets have their purpose but if they are made from cotton or natural fibers then pesticides or herbicides may be found in the fibers. Since there is no way to know this you may want to choose a synthetic weave. In any case you will want a shallow weave since that will hold less dust and make it easy to care for. Carpet and upholstery stains can be removed with club soda.

Hardwood floors are attractive but expensive. They have to be finished and these finishes can give off a very high level of toxic volatile compounds unless you are careful in your selection of care products. The floors scratch easily but damp mopping will work. The volatile compound level could reach 100 times the normal indoor level depending on how much hardwood you have and how often the floor is treated and waxed.

Maintaining a good hardwood floor can put your respiratory health at risk. This is because it must be waxed and treated on a regular basis, in most cases. In this process, a very high level of VOCs can be achieved, all to your detriment.

Other less popular and more expensive types of floor covering include stone, terrazzo, quarry tile and cork.

CASE HISTORIES

CASE NO. 1: A home has a leaky roof and the ceiling is wet. The wet spot moves toward a wall where it stops. The occupant complains that there is the smell of mold in the home. The odor is heavy and it is annoying. The home is monitored and mold spores are not found in the air. In this instance the water has run down behind the wallpaper and mold growth has occurred there. The wallpaper has prevented the spores from entering the room, but it has not prevented the smell of mold from penetrating the paper. In this case one can find the mold by looking for soft spots and bubbling in the wallpaper.

With a small sharp knife cut out three sides of a rectangle about one-half inch on a side and peel it back. If there is mold present you will see its black or green appearance on the wall and on the back of the paper. If not, carefully replace the paper and repair it later. Continue the inspection.

Once you define the limits of the contamination obtain some sandpaper and prepare a solution of baking soda or warm soapy water (one-half cup of baking soda per bucket of warm water). Carefully peel off the wallpaper, all the while wetting the paper with a wet sponge. This is done to prevent spores from entering the room. Then wet-sandpaper the wall and discard the wallpaper that you removed into a plastic garbage bag. Dry the wall after the dark stains have been removed and you are now ready to re-paper or re-paint.

CASE NO. 2: A married couple buy an older home and one of them complains of headaches and occasional nausea upon awakening each morning. Most of the chemicals have long since out-gassed from the home. There have been no roof leaks or serious water damage and the home is in a low outdoor allergen area.

In this case, the home has a double car garage. When a car is driven into the garage at night, the garage door is closed, thus sealing in the exhaust fumes and the greases from the heated engine. One of the couple enters the home by a door that adjoins the kitchen and keeps the door open to bring in groceries. The fumes enter the home and are picked up by the return air duct of the air handling system, which is located indoors and near the garage. The fumes are recirculated throughout the home, which includes the bedroom. Vapors from the garage also leach through the walls of the garage into the interior spaces of the home.

A good year-round compromise solution is to have a turbine fan in the roof of the garage or an electrically operated fan to exhaust the air of the garage. Remember that this air has to be replaced from somewhere, so you will need to create a way for the air to enter the garage from the opposite side for flow-through purposes.

CASE NO. 3: A retired couple have been coming to their second home each winter home for the past twenty years. Their son has diligently maintained the home in their absence. The couple noticed that for the past five years each of them develops a cough within days after their arrival and the cough stays with them until after returning to their primary residence. The gas company says that the furnace is good, they have no pets, they park the car away from the house, neither has any allergies and they live a simple life.

Under the microscope, samples of air from the home reveal high quantities of fiberglass spears bound with yellowish urea-formaldehyde. Pieces of glass the size of pollen grains, and mold spores, are also present. Further investigation of the home reveals an attic with old style fiberglass cotton-like flocking that was used as insulation.

The heater and ducts are located in the attic. The duct is found to have a split at its juncture with the heater. This serves as a vacuum when air is forced through the duct and the fiberglass is not only sucked into the duct, but is sheared off to create even smaller, sharper fragments that are inhaled.

To remedy the problem the duct had to be taped and the house had to be thoroughly damp-dusted and vacuumed. The vacuuming had to be done with the windows open and the vacuum operator had to wear a mask. Once these tasks were completed the couple ceased to have respiratory problems in this home.

CASE NO. 4: The owners of a very old mobile home do their best to maintain it. The home has leaked from time to time, is evaporative cooled in the spring and summer months and formaldehyde is no longer a problem as far as off-gassing from structural textiles from within the home.

During the summer the elderly father develops a cough, but his room has a window air conditioner. This unit is found to be draining properly away from the home. The evaporative cooler is suspected of transmitting bacteria and mold into the home but investigation clears this unit. The father's room, however, is found to have a high level of tiny Aspergillus mold spores. Once the filter on his air conditioner unit is thoroughly cleaned his cough disappears.

CASE N0. 5: A woman lives alone and has allergies to tree pollen every spring. Every April the problem becomes severe and she is reactive day and night. Her home is found to have a high level of elm pollen indoors; an elm tree is located by her bedroom window and she states that she likes to keep the bedroom window open because she likes fresh air. It is diplomatically suggested that she try to open some windows on the opposite side of the house to se how she feels.

APARTMENT LIVING

There are many ways that cigarette smoke or cooking odors or indoor pesticides can travel on airways from one apartment unit to another. Whether you are the giver or the receiver, the same will apply to you.

In general it is safe to say that odors are gases and can mix with air. As such they can go anywhere that air can go. Any break in the wall can be an outlet or an inlet for these odors. What can the average person do at little cost with or without the permission of the management.

There are a lot of pressure changes between your unit and the next door, above or below unit, depending on whose air conditioner or heater is on at the time. This will account for smells that occur at odd hours of the day or night.

Use tape to seal the places under the sinks or any area where pipes go into the walls. It also includes pipes near the water heater. If you live in a townhome or condominium, this includes the hole where the dryer hose goes through.

Aluminum tape is thinner and seals better than duct tape and it lasts longer without drying out. It is also more resistant to weather. Naturally it is going to cost more.

Prices will vary but you should be able to buy foil tape at no more than fifty percent more than duct tape.

Keep in mind that there is air between the walls. Electrical outlets and switches permit smells to leave or enter an apartment. Check at hardware stores for electric outlet seals for receptacles and light switches. The cost is slight and what you get is foam switch seals for your on-off switches, foam receptacle seals for your plug-ins, and safety caps to plug into unused outlets. You can unscrew the plastic switch plates yourself and just insert these items behind them.

Take off the registers and use silicone to seal the space where the ducts are not flush against the walls. You can also use it to seal the area where lamps hang from the ceiling. Most silicone is made with acetic acid, otherwise known as vinegar, so its smell may or may not be annoying. The good news is that it will disappear in a short period of time.

Silicone costs between four and five dollars for a 10 ounce tube and comes in different colors for different purposes.

CHAPTER 9
BUYING AND SELLING YOUR HOME

More and more realtors are becoming aware of the allergy / asthma / chemical sensitivity issue. More and more persons are expressing their desire to find a clean and safe residence for themselves and their children. Therefore, the advice of realtors to owners can go a long way toward preparing the home for sale.

From another standpoint, there is a lot that the homeowner can do to get the home as clean and fresh as possible prior to sale. Too many homes and rentals are passed by because the prior inhabitants meant well, but did the wrong things.

If you are planning to sell or lease the home take a look around you. Neat and orderly is nice, but what do you smell in the various rooms? Does the bathroom(s) reek of perfume, aftershave, chlorine bleach and pine-scented toilet bowl cleaner? Is the toilet paper and tissue fragranced? Do you have stinky cleaners and polishes under the sink? What's in the medicine cabinet? Whether you like the smells or not is not the issue. We are talking about fresh unscented air because for a lot of people certain smells are a red flag and no smell at all can mean a green light.

Check out your laundry area, storage area and garage for smells. The garage floor should be scrubbed down with a solution of five percent trisodium phosphate (TSP) (about 2 oz. per quart of water. TSP is odor free and will remove the grease and attendant odors. It is extremely inexpensive and is available in hardware stores.

Wash the floors and walls and counter tops with a solution of baking soda. For this you can use a half cup of baking soda in a bucket of warm water. You will be surprised how inexpensive and effective it can be.

If you are planning to clean the home and then go away to leave it in the hands of a realtor to lease or sell, well and good. Just do not have the carpets cleaned and shampooed and then lock up the house. Allow time to make sure the carpets are dry, otherwise residual moisture from the carpets will aerosolize any smells that the carpets have absorbed. These homes frequently smell badly when they are opened (sometimes weeks and months later) and it is very difficult to remove the odor at that time because it has now become part of the walls and ceiling. From the realtors' standpoint idle homes need to be completely opened and aired out on occasion anyway, and not just when there is a showing. Then it is too late.

The air conditioning or heating filter should be changed prior to showing in case some mechanically minded person wants to look at it. Besides, the presence of clean filters suggests that there is care for that home. Also, for desert dwellers, the evaporative cooler should be cleaned and serviced prior to sale so that it is quiet and works efficiently. Adjust the air handling registers or vents so that they do not blow on your head. You may be used to it but it is a turnoff for a prospective buyer to be standing in your kitchen and have cold air, hot air or gale force winds blowing on their head and face.

People with respiratory disorders will shy away from a chemically sanitized home and includes half of our population.

People who want to rent or buy a "chemically free" household find that there is no such thing. In fact, every home has some chemical-type problem associated with it. If you are planning to sell or lease your home here are some tips that will help improve your chances of finding a caring client.

First, some general rules: Do not use air fresheners to cover up odors, and avoid use of scented products for cleaning purposes, if at all possible. Many people think that if a dab of perfume is nice then a lot more is a lot nicer. Generally, the smell of perfume cannot be removed from a home.

If your home has a garage, make sure that you clean the garage thoroughly. This means scrubbing the floors with a solution of trisodium phosphate (TSP), as noted before. Also, clean out the cabinets in the garage so the potential buyers can see how much space they will have. Wear gloves and use the TSP solution in the cabinets. Be aware that the TSP will remove some of the gloss from your varnished surfaces. That's okay. Let the next buyer varnish them if he chooses to do so.

The reason you want the garage clean is because vehicle odors migrate into the home. They are absorbed into the walls and readily enter the home via cracks around the door attached to the home. They also enter the return air duct to be circulated throughout the home. Make sure the door to the home is well sealed.

The laundry room should be clean of old soap residue on the machines and floor. Wash the shelves with your TSP solution (use gloves) to remove the smells of chlorine bleach, detergent, pine scent, fabric softeners, antistatic products, paint thinner, shoe polish, furniture polish and pesticides. This area is a hot spot where out-gassing occurs into the rest of the home and is sure to annoy some potential buyer.

A solution of baking soda will serve as a substitute for TSP in less severe locations that need cleaning.

If you are one of the many who have their indoor baseboards sprayed with pesticide once a month, cease and desist. Open the home to the sunlight and fresh air and let this material degrade as quickly as possible. Wash the baseboards with a scrubber of your choice and warm boric acid, or again, TSP. Many commercial products are odor free and cannot be detected.

Open the home to ventilate while you are doing this chore. Unless the pesticide is some odor-free product it will carry into the air of the home. Check with the people who provide this service for you and find out what they are using and how to get rid of it.

If possible, keep your pets out of the home when potential buyers are present. People who are concerned about allergies do not like to see pet food, pet toys or animals running around and jumping on them. They are looking for a home that is going to be good to them.

The ideal home has a very low background level of smells. Home smells include everything from cooking odors to petroleum-based hydrocarbons, formaldehyde, soaps, deodorants, cleaners, paints, freshly brewed coffee, microwave popcorn, laundered and dry-cleaned clothing, and a variety of other sources. If you just sprayed the home with insecticide or poured chlorine bleach or pine scent down the toilet, the level of smells can be much higher than normal. The garage frequently, but not always, measures the highest hydrocarbon level in the home with readings of up to and over seven parts per million (ppm).

The use of too much wood polish can raise the concentration of volatile organic compounds (VOCs) to an irritating level of as much as fifteen ppm. This can kill a sale.

This background level should be as low as possible when the realtor and prospective buyer walks in. Make sure the home is clean and dry before showing it.

In general, carpets are becoming an increasing problem, from the out-gassing of new carpeting when improperly cleaned, to the smells that are absorbed into older ones. If there is tile or wood flooring beneath your carpeting you might consider just removing a roomful of the carpet to expose the plain floor.

Obviously the yard and the alley need to be kept trim to reduce the incidence of allergenic pollen from weeds and grass. If you have a lawn it is a no-win situation for an allergic or asthmatic person out of doors and as well as indoors due to tracking of particles into the home.

OUTDOOR FACTORS

In many cities and communities one area is similar to another. But exposure to allergens and irritants can still be minimized with a little homework. First, it will be very helpful for you to know the prevailing wind direction during the spring and fall months. Communication with your local weather service, radio or television newscaster, or just awareness and common sense can all be of help. Realize, however, that the wind direction in a specific locale can be very different than the official reports due to the presence of mountains or man-made structures.

Next, it is helpful to know which of the three you are sensitive to: pollen, mold or dust. This knowledge will help determine what side of the city you may want to live on. For example suppose you are sensitive to tree pollen of all types and the prevailing springtime breezes are from the south, then you may want to find a residence on the south side (upwind from the trees of your city), if that is possible. If you are sensitive predominantly to fall weed and grass pollen and the fall breezes are from the north then you may want to reside on the north end of the city, unless there is a weedy lot to the north of the home.

Lower parts of the city and cooler temperatures in the morning hours bring particles in the atmosphere downward to settle out. Thus, areas that are significantly lower in elevation should be avoided.

The residence itself should have a minimum of vegetation around it to reduce the amount of local mold and pollen. Each home has a microbial shell around it that is influenced primarily by the local vegetation., secondarily by neighboring vegetation and finally by distant vegetation. Pay attention to the type of landscaping that your neighbors have. If there is a lawn associated with the residence expect your exposure to a wide variety of allergenic grass particles and mold to be elevated. Incidentally, that includes a well kept lawn. Who is going to cut the grass?

Learn your trees. Watch for trees that are too close to windows of the residence. Or, if trees are next to a main walkway, they would shed pollen that could be tracked indoors. In either case, when the trees pollinate in the spring you may have an indoor pollen problem.

You may also wish to find out if any major road construction will be taking place in your prospective new area since this means tar fumes, diesel smoke and dust outdoors, and possibly indoors, for a long period of time. You city planning and zoning department can help you here. Do not rely on your realtor for this information since this is not his or her area of expertise.

Look over the back fence of the residence. Is the alley clean or full of grass and weeds that nobody except you will ever cut?

Find out the average amount of rainfall for the city you wish to move to. Amounts of aeroallergens are generally related to rainfall since more rainfall means more plants, more pollen and more mold.

If you are looking at apartment units, watch for piles of swirling dust on certain doorsteps. This is a clue that channel drafting is occurring and that the residence is at the end of a natural or man-made channel (such a between buildings). Wind will travel down these channels carrying debris. This will cause a much higher rate of build-up of allergenic particles within the building and increase your exposure level.

There is evidence that any kind of windbreak will significantly reduce the exposure of your residence to allergenic substances being carried on the wind. This means that stand of oleander or insect pollinated trees, a fence or other building, can help redirect the flow or particles away from you.

No residence or area is going to be perfect. Be prepared to make tradeoffs.

INDOOR FACTORS

Proper weather stripping has its good and bad sides. On the plus side, it keeps out dust and aeroallergens. On the minus side it does not permit the home to breathe. A home needs to be ventilated with fresh air to reduce the build-up of a variety of indoor pollutants such as outgassed materials, sweat and cooking odors.

The level of indoor humidity will vary greatly daily and seasonally and will have to be controlled by the use of a dehumidifier or other methods in order to keep down the level of molds and mites. The degree of weather-proofing you do will depend on your level of sensitivity to outdoor or indoor pollutants. You can check with your local home builders'association for recommendations in this regard.

Look for fan ventilation in the bathroom and kitchen to minimize moisture build-up of mold in the bathroom (usually on the shower ceiling in unvented bathrooms) and mites in the rest of the home.

Ask if there have been any water leaks indoors, usually caused by broken pipes or leaking water heaters. If the carpets and base boards have not been thoroughly dried within three to four days after the break there may be a real mold problem caused by any number of mold species such as Penicillium, Stachybotrys, Phoma, Cladosporium, and even slime molds, all of which are potentially allergenic species to your family, house guests and pets. Do the present or previous owners have a coughing problems when indoors?

When you first inspect your potential new home or apartment look for stains on the ceiling, walls, floor and carpet and try to find out what caused them. Has the roof leaked? Is the bathroom unvented? The presence of ceiling stains does not necessarily mean that mold growth is present unless the mold growth is obvious.

Does the residence have an attic with duct work? If so, there is an even chance that allergens in the attic or hazardous particles such as fiberglass located there are gaining entrance into the home proper.

Check out the laundry room. If you are annoyed by fragrances and chemical smells then you are going to have to scrub out this area when you move in.

Does the refrigerator come with the home? Then it may have to be inspected and decontaminated.

Check the condition of the fireplace. A door that closes poorly means that fumes and soot will escape into the home.

What kind of filters are being used in the central air handling system? If they are the standard fiberglass variety you will have to upgrade to a pleated filter.

Note the location of the garage and the entry into the home. It a return air duct located by this entry? This could be a source of auto exhaust into the entire home. You may need to thoroughly weather strip this door.

Homes that have shade on all sides have potential for mold build-up. Try to find a home that has a reasonable amount of exposure to natural light.

A great deal of indoor pollen and spores occur by tracking. This is especially true when trees are located near a walkway or there is a yard that needs mowing. An outdoor mat made of rough fiber located by the entrance to the front and back door can go a long way toward lessening this potential problem. Some persons take off their shoes upon entering the home.

CHAPTER 10
POLLEN AND MOLD DATA

The information presented below is meant to be a general guideline for aeroallergens for cities in the United States. These data were obtained from numerous sources including publications from the American Academy of Allergy, Asthma and Immunology and from numerous scientific papers. No attempt was made to review all of the scientific literature; therefore, this table is intended to serve in the most general comparative sense.

The numbers presented are rounded off for easy reading. These figures depend upon the number of monitors available to a given city and can vary greatly from one area of the city to another and from one year to the next.

Only the highest pollen and mold counts are presented in terms of number of particles per cubic meter per day. Because of the complicated names of molds abbreviations are presented in most cases: Clado = Cladosporium (Hormodendrum), Alt = Alternaria, Helmin = Helminthosporium (now known as Dreschlera), Fus = Fusarium, Asperg = Aspergillus, Penicil = Penicillium.

Numerous pollens and molds were omitted for the sake of simplicity. Virtually all of the cities listed were done so because the type of monitor was the same. Numerous cities were omitted because the type of monitor was different and the information could not be compared with the cities listed below.

Albuquerque, N. Mex.

POLLEN	MONTH	AMT	MOLD	MONTH	AMT
juniper	May	500	Clado	Nov	22000
ash	Apr	2500	Helmin	June	300

Austin, Tex.

juniper	Jan	2700	Clado	May	12000
oak	Apr	2500	Alt	Aug	8600
elm	Sept	3000	Asperg	Sept	3200
ragweed	Oct	2500			

Billings, Mont.

poplar	Apr	350	Clado	June	1500
grass	June	200	smuts	July	2300
sagebrush	Sept	100	Helm	Aug	100

Boise, Idaho

sycamore	May	450	no information available		
sagebrush	Oct	550			

Boulder, Colo.

poplar	May	2800	Clado	Aug	12000
grasses	June	2600			
ragweed	Aug	2500			

Brooklyn, N.Y.

oak	May	3100	Clado	Aug-Sept	46000
sycamore	May	2000	Pullularia	July	1700
hackberry	May	900	Fusarium	Aug	1600
Mushroom spores	Aug-Sept.	7000			

Burlington, Vt.

POLLEN	MONTH	AMT	MOLD	MONTH	AMT
birch	May	2400	Clado	Sept	500
ragweed	Aug	350	Mushroom Spores	Sept.	33000

Cape Girardeau, Mo.

POLLEN	MONTH	AMT	MOLD	MONTH	AMT
oak	Apr	14000	Alt	Sept	13000
elm	Mar	6100	Clado	May	5000
maple	Apr	2200			
ragweed	Sept	6400			

Clayton, Mo.

POLLEN	MONTH	AMT	MOLD	MONTH	AMT
ragweed	Aug	2500	Clado	Aug	14000
grasses	June	900	Mushroom spores	Aug	25000

Colorado Springs, CO

POLLEN	MONTH	AMT	MOLD	MONTH	AMT
elm	May	26000	no information available		
poplar	May	11000			
ragweed	Aug	9500			
grasses	Aug	2300			

Daytona Beach, Fla.

POLLEN	MONTH	AMT	MOLD	MONTH	AMT
juniper	Feb	80	no information available		
oak	Apr	80			

Des Moines, Iowa

POLLEN	MONTH	AMT	MOLD	MONTH	AMT
ragweed	Aug	9800	Clado	Aug-Sept	27000
nettle	Aug	1100	rusts	Sept	1100
grasses	June	750			

El Paso, Tex.

POLLEN	MONTH	AMT	MOLD	MONTH	AMT
mulberry	Mar	5700	rusts	Aug	1100
juniper	Mar	1100	Alt	Aug	700

Fresno, Calif.

POLLEN	MONTH	AMT	MOLD	MONTH	AMT
olive	May	400	Clado	Sept	6400
grasses	June	900	smuts-rusts	June	6000

Kalamazoo, Mich.

elm	Mar	125	no information available		
mulberry	May	25			
ragweed	Aug	140			

La Jolla, Calif.

oak	Mar	250	Clado	Sept	11000
grasses	Aug	225			

Mason City, Iowa

maple	May	150	Clado	Aug-Oct	160000
ragweed	Aug	3700	smuts	Oct	4600
grass	June	1100			

Minneapolis, Minn.

Oak	May	1500	Clado	July-Aug	12000
juniper	Apr	1500			
oak	May	1500			
ragweed	Aug	2300			

Newark, N.J.

birch	May	250	no information available		
oak	May	250			
ragweed	Sept	50			

Portland, Maine

birch	May	2100	mushroom spores:	July	1900
oak	May	1800			
ragweed	Aug	61			

Providence, R.I.

POLLEN	MONTH	AMT	MOLD	MONTH	AMT
oak	May	26000	Clado	Aug	11000
birch	May	10000	Fusarium	Aug	1500
ragweed	Sept	4200	Curvularia:	July-Aug	1100

Reno, Nev.

juniper	Mar	50	no information available		
elm	Mar	50			
sagebrush	Oct	40			

Sacramento, Calif.

mulberry	Mar	12000	Clado	Apr	9000
birch	Mar	200	Alt	Mar-Apr	1500
alder	Jan	150			
grass	Mar	1500			

San Diego, Calif.

oak	Apr	100	no information available		
mulberry	Apr	20			
grasses	June	15			

San Francisco, Calif.

juniper	Mar	350	Clado	Nov	70
chestnut	May	100	Alt	May	50
grass	May	400			

Saratota, Fla.

oak	Mar	2300	Clado	Sept	7600
Aust.pine	Oct	300	Helmin	July	800
palm	June	100			
grasses	May-Aug	600			

Spokane, Wash.

POLLEN	MONTH	AMT	MOLD	MONTH	AMT
pine	June	4600	smuts	June	7200
grasses	June	600	Clado	Aug	1600

Tucson, Ariz.

mulberry	Mar	250	Clado	Oct	150
olive	Apr	100	Alt	Oct	50
ragweed	Mar	25	smuts	May	35
grasses	Sept	25			

Tulsa, Okla.

elm	Mar	2500	Clado	Sept	12000
oak	Apr	2100	Penicil.	June	800
poplar	Apr	1900	mushroom spores:	May	3200
ragweed	Sept	11000			
grasses	Apr	1500			

Washington, D.C.

hickory	May	750	Alt	July-Aug	300
birch	May	650	Fusarium	Aug	300
oak	May	300			
grass	June	200			

Additional notes: Juniper also includes cypress and cedar; pine is not listed for space considerations and because it is minimally allergenic; when two or more months are listed this means that the counts are about the same for each of the months.

SPECIAL ESSAY:

DOES TELEVISION CAUSE ASTHMA?

There are many positive aspects to television viewing. It provides entertainment for the masses, and especially those who are elderly or infirm.

As with most things, there are basically three schools of thought regarding television-viewing practices. There are those who do not advocate watching TV at all. There are those who advocate watching in moderation and with careful selectivity, and finally those who don't care and watch to their heart's content.

In Russia, not too many years ago, there was a saying that one had a choice of two channels, the government sponsored station, or the set could be turned off.

This essay seeks the moderate point of view, through the condemnation of television. That is to say, by promulgating the negative aspects of TV watching, the average person may have a tendency to back off. This is a good thing.

For decades the rate of asthma has been increasing in industrialized nations around the world. This includes the disease among children, as well as adults. There have been a lot of theories regarding this phenomenon, but no hard answers, as yet.

Air pollution is not the primary reason for the problem, because the asthma rate still increases in cities where the level of pollution has actually decreased.

Indoor air pollution is a real possibility, since the number of hazardous home products has increased along with the population. Certainly, if we look at the number of fragrance products, pesticides, cleaning solvents, polishes, personal care products and volatile organic compounds in general that have entered our indoor air space over the decades, the truth may be in this realm.

Another possibility is that television acts as the catalyst in the presence of indoor pollutants, or it may act independently to a lesser degree without their presence.

Since the biggest asthma trigger is stress, the reasons I present below regard the conversion of the healthy person to the asthmatic in the presence of stress——stress caused by television.

Reason No. 1: The average American watches some five hours of TV per day. Over the period of a year the total comes to 1825 hours. That computes to 228 8-hour days; enough time to learn several new languages, learn a completely new trade, write a novel, possibly read the complete works of Shakespeare or Josephus, learn the fundamentals of rocket science or ornithology or study for some ten college courses. That's per person per year.

Television eats up our free time so we have less time to spend with the family, eat a decent meal, go on vacation, exercise, or to earn extra income.

Parents spend an average of 38.5 minutes per week of meaningful conversation with their children. The average American child spends 2100 minutes per week watching television.

Now, out of those 1825 hours of TV watched per year, approximately half that time is commercial related, in one form or another. We watch someone yell at us with the volume increased, point their finger, trick, induce, and sucker us. Occasionally, the honest person appears, but requires a gimmick to be competitive. Blatant lies have become the in-thing. Witness a President who sets the example for the nation.

This isn't complicated. It's called stress.

Reason No. 2: Want more stress? Try watching the news as the media hypes the smallest issue into a national event. We are enticed to follow the sponsored coverage and to come right back after the break. In the end, the event provides a format upon which even our dreams are based.

Television has created a generation of adults and their children who lack imagination. The reason for this is that television presents the hype——a stressful exaggeration of events to hook and control the viewer. People in real life become apathetic and an apathetic person does not strive to become better. If we can't fathom the difference, then we are unquestionably a victim of our times. Low voter turnout is only one example of our apathy. Refusal to become angered at state and federal fiscal abuses is another example.

At one time, it was thought that a world war was necessary to bring us out of our lethargy. I offer that it just takes more and better writers and speakers. Yet these persons, as well as classical composers and great painters, cannot be born anew in the era of the television; the desire to create something new is a rare desire. Just as necessity is the mother of invention, so boredom is the mother of creativity. Television will not permit us to be bored.

With great sadness I submit that television will not bow out with a curtsey, with dignity or self esteem intact, and exit the stage.Therefore, the task befalls us, as readers and as listeners, to become as great as we can be.

Reason No. 3: Spending money on the products that we want, but don't need, forces us into economic stress. We are in debt as a nation because we are in debt as individuals. We wouldn't let that yelling person walk in our front door, but we watch them hour after hour through television. Why do we do this? Let the psychologists answer that one.

By the time he or she is sixty-five years of age, our American will have viewed two million commercials and will have spent nine years of life (*nine years)* staring at flickering and artificial light.

Americans are not unusual in their television viewing habits. These figures are actually low for some countries, where for tens of millions of households, television is an addiction used by the rich and the poor. In fact, television is worse than addiction to drugs, alcohol and cigarettes. This is because over a billion people are addicted to it worldwide.

Reason No. 4: Images fill our mind, as Jerry Mander so eloquently stated in his book, "Four Arguments for the Elimination of Television."

They are not images of creative thinking or original thought. They are somebody else's images. TV is not relaxing. It is just the opposite. If it were relaxing we would be able to sleep well, after watching hours of it each evening before bed. Poor sleep patterns mean more stress.

I wonder where videogames and the Internet chat rooms fit into this picture.

Reason No. 5: Watching violence and action leads to violent and active thoughts and behavior. This has been well researched and documented for decades. Ask an experienced teacher if their students have less ambition these days compared with twenty or even ten or even five years ago.

By the time today's American is 18 years he will have viewed 200,000 acts of violence on TV. One can only imagine a perfect world where this many acts of kindness are performed instead.

Television viewing is full of conflicts and over-stimulation. Neither the child nor the adult is able to release the physical stress that they feel, due to this over-stimulation. Conflict resolution by this medium is miniscule compared with the stress induced upon the viewer.

Reason No. 6: a) An immune system that operates on a diet of TV dinners and fast food is not in a position to react well to allergens, b) The nervous system requires exercise, c) There is no requirement for artificial lighting to enter the eyes.

For hours each day the television viewers don't see sunlight and don't read a book. Our next generation of leaders may not be able to read or write. Today's students can't. Why should it change for the better? That increases the stress load that we feel; more stress for those job seekers who lack fundamental skills of mathematics and varied aspects of linguistics, the two skills most closely correlated with success in life.

It has always interested me that we complain about incompetent teachers, while both the parent and their children languish in front of the television. I think that we find excuses to indulge in this, our habit.

Reason No. 7: Some eighty-five percent of asthma is allergic in nature. Studies have shown us that indoor air is much more polluted than outdoor air. Since we spend more than ninety percent of our time indoors, we surround ourselves with toxic products, which have been sold to us through advertising on television. I mentioned in the previous articles that there is no regulatory mechanism to screen the products that are sold for in-home use.

We breathe gases produced from structural materials, glues, solvents, paints and dry cleaning chemicals. The products in our homes fill the air with petroleum distillates, secret 'inert ingredients'. The list is a lengthy one.

Each of these is a known respiratory irritant and asthma trigger, which may have its effect in the early morning hours when the daily hormone level reaches a peak and when many asthma attacks occur.

And then, before we leave the bathroom in the morning we have exposed ourselves to some twenty different products with fragrance. This list includes toothpaste, mouthwash, shampoo, bar soap, conditioner, deodorant, body talc, hair spray, facial and toilet tissue, a wide variety of cosmetics, perfume and aftershave and clothing that has been laundered in fragrance detergent and dried with fragrance anti-static products. All of this occurs with little or no fresh air to dilute the mix.

A cruel synergism occurs, an enhancing effect that television has upon the viewer whose immune system is already assaulted by stress in the form of products. The end result is a worldwide increase in asthma in industrialized nations, those same nations that "modernize" their product lines and sell televisions to the masses.

This television is the black hole in the home. It sucks up energy, time, money, creativity, vitality, and family life.

It returns transient and shallow entertainment and a few educational programs, if the viewer knows where to look. Television gives back stress, obesity, sleeplessness, cultural depravity, bad grades in school, a violent society and asthma attacks that result the stress load.

As for myself, I am not going to quit television cold turkey. But these days, the mute button is at hand and some project is in my lap, even as I cut down on viewing time.

Oh, I also believe that television is responsible for the increase in the cancer rate, for many of the same reasons, but that's another story.

SELECTED BOOKS

1. Clean and Green, by Annie Berthold-Bond, Ceres Press. 1990.

2. The Non-Toxic home, by Debra Lynn Dadd, St. Martin's Press. 1986.

3. Clean House, Clean Planet, by Karen Logan, Pocket Books. 1997.

4. The Healthy Home, by Linda Mason Hunter, Pocket Books. 1989 ($9.95).

5. The Consumer's Brand-Name Guide to Household Products, by Carol Ann Rinzler, Lippincott and Crowell. 1980.

6. Allergic to the Twentieth Century, by Peter Radetsky, Little, Brown and Co. 1997.

7. Your Home, Your health and Well-Being, by Rousseau, Rea and Enuright, Hartley and Marks. 1988.

8. Nontoxic, Natural and Earthwise, by Debra Lynn dadd, Jeremy P. Parcher, Inc. 1990.

9. Is This Your Child's World? How you can fix the schools and homes that are making your children sick, by Doris J. Rapp, M.D., Bantam Books.1996.

10. Indoor Pollution, by Steve Coffel and Karyn Feiden., Fawcett Columbine. 1990.

11. Allergy Free, by Scott E. Seargeant, Seargeant Publ. Co., Inc., 1997.

12 . Baking Soda. Over 500 Fabulous Fun and Frugal Uses You've Probably Never Thought Of, by Vicki Lansky, Book Peddlers .1995.

13. Toxic Deception, by Dan Fagin, Marianne Lavelle, and the Center for Public Integrity.Carol Publ. 1996.

14. A Consumer's Dictionary of Cosmetic Ingredients, by Ruth Winter, M.S., Three Rivers Press. 1994.

15. The Healthy House. How to Buy One, How to Cure a Sick One, by John Bower, Barol Communications. 1989.

16. Healthy House Building...A Design and Construction Guide, by John Bower, The Healthy House Institute. 1997.

17. Common-Sense Pest Control...Least Toxic Solutions for Your Home, Garden, Pets and Community, by Olkowski, Daar and Olkowski. Taunton Press. 1993.

18. Work is Dangerous to Your Health, by Jeanne M. Stellman, M.D. and Susan M. Daun, M.D., Vintage Books. 1990.

19. Toxics A to Z. A Guide to Everyday Pollution Hazards, by Harte, Holdren, Schneider and Shirley. Univ. of California Press. 1991.

20. Chemical Exposures—Low Levels and High Stakes, by Nicholas Ashford and Claudia Miller. Van Norstrand Reinhold. 1991.

21. Indoor Air Pollution and Health, Bardana Jr., E. and Montanaro A.eds., Marcel Decker, Inc. 1997.

22. Your Home, Your Health, and Well-Being, by D. Rousseau, Rea WJ and Enright, J. Hartley and Marks. 1988.

23. Fast Food Nation: The Dark Side of the All American Meal, by Eric Schlosser. Houghton Mifflin. 2001.

24. Four Arguments for the Elimination of Television, by Jerry Mander, Quill. 1978.

25. For more information on carpeting you may write The Carpet and Rug Institute, P.O. Box 2048, Dalton, Georgia 30722 (1-706-278-3176); the U.S. Consumer Product Safety Commission, Washington, D.C. 20207 (1-800-638-2772); and the U.S.Environmental Protection Agency, TSCA Assistance Information Service (TS-799), 401 M Street, S.W., Washington, DC 20460 (1-202-554-1404); and finally the Technical Bulletin on Carpet and Indoor Air Quality in Schools, Maryland State Dept. of Education, Division of Business Services, 200 West Baltimore Street, Baltimore, Maryland 21201 (1-410-333-2508) or ask the EPA for a free copy of this bulletin.

26. If you would like to make your view heard regarding fragrances, you may wish to contact U.S. Congressman Ron Wyden, Cosmetic Safety, 1406 Longworth H.O.B., Washington, DC 20515. (Congressman Wyden is chairman of a congressional subcommittee that is investigating cosmetic safety.) A free copy of "Neurotoxins at Home and

in the Workplace" is available. This is a congressional report. Phone 202/225-4494 or write to Rayburn H.O.B., Washington, DC 20510. Ask for report 99-827. Also, If you are sensitive to fragrances write to the Food and Drug Administration, Attn: Heinz J. Eiermann, Director, Division of Colors and Cosmetics, Washington, D.C. 20204.

27. Information about aroma therapy is readily available on the Internet. The interested reader is encouraged to read more about the subject. Some of the newer publications include: Aroma therapy Workbook, by Marcel Lavabre,$12.95; Aroma therapy Handbook, by Erich Keller, $12.95; The Practice of Aroma therapy, by Jean Valnet, M.D., $10.95. All three books are published by Healing Arts Press.

28. I would like to thank the National Center for Environmental Health Strategies, 1100 Rural Avenue, Voorhees, New Jersey 08043, for providing me with a great deal of information about fragrances and chemical sensitivity.

VIDEOS

1. Environmentally Sick Schools, Students and Teachers at Risk, by Doris J. Rapp, M.D., Practical Allergy Research Foundation. (Call 1-800-787-8780) 90 minutes

2. Coping with Allergies: Your Guide to Quick Relief From Sneezing,Coughing and Congestion, by Xenejenex, The Health Care Communications Companies; Xenegenex Health Videos. Call 1-800-228-2495.

SELECTED SCIENTIFIC PAPERS

Key to journal abbreviations:

Acad	academy	Immunol	immunology
Allerg	allergy	Ind	industry
Amer	American	Int	internal
Ann	annals	J	journal
Arch	archives	Med	medicine
Chem	chemical	Occup	occupational
Clin	clinical	Onc	oncology
Contam	contamination	Proc	proceedings
Emer	emergency	Resp	respiratory
Engin	engineering	Rev	review
Env	environmental	Tox	toxicology
Exp	experimental	Persp	perspectives
Expos	exposure		

Asthma and the home environment. Review Article. Jones AP. J Asthma. 2000, 37(2):103.

Effects of allergy season on mood and cognitive function. Marshall PS and Colon EA, M.D. Ann Allerg. 1993, 71:251.

Global increases in allergic respiratory disease: The possible role of diesel exhaust particles. Peterson B and Saxon A. Ann Allerg Asth Immunol. 1996, 77:263.

Airborne house dust particles and diesel exhaust particles as allergen carriers. Ormstad, H, et al. Clin Exp Allerg. 1998, 28:702.

Asthma in United States Olympic athletes who participated in the 1996 summer games. Weiller JM, Layton T and Hunt M. J Allerg Clin immunol. 1998, 102:722.

A health hazard assessment in school arts and crafts. Lu P. J Env Path Tox Onc. 1992, 11:12.

Learning impairment and allergic rhinitis. Simmon F, M.D. Allerg Asthma Proc. 1996, 17:4.

Indoor allergens and asthma: Report of the third international workshop. Platts-Mills T, et al. J Allerg Clin Immunol. 1997, 100(6):S2

Effect of odors in asthma. Shim, C, M.D. and Williams, Jr. MH, M.D. Amer J Med. 1986, 80:18.

The effects of fragrances on respiratory reactions of asthmatics. McCants M., et al. Amer Acad of Allerg, Asthma Immunol. 56th Annual Meeting, Abstract No. 371.

Occupational asthma caused by aromatic herbs. Lemier C, et al. Allerg. 1996, 51:647.

Prevalence of dust mites in the homes of people with asthma living in eight different geographic areas of the United States. Arlian L, et al. J Allerg Clin Immunol. 1992, 90(3):292.

Dust mites: Immunology, allergic disease, and environmental control. Platts-Mills T., M.D. and Chapman M., M.D. J Allerg Clin Immunol. 1987, 80(6):755.

Dog allergy: Understanding our 'best friend'? Tubiolo V.C., and Beall GN. Clin Exp Allerg. 1997, 27:354.
Breed-specific dog-dandruff allergens. Lindgren, S., M.D., et al. J Allerg Clin Immunol 1988, 82(2):196.

Cat antigen in homes with and without cats may induce allergic symptoms. Bollinger ME, D.O., et al. J Allerg Clin Immunol. 1996, 97(4):907.

Carpet properties that affect the retention of cat allergen. Lewis RD and Breysse PN. Ann Allerg Asthma Immunol. 2000, 84:31.

Allergens of horse dander: Comparison among breeds and individual animals by immunoblotting. Felix K., et al. J Allerg Clin Immunol. 1996, 98(1):169.

Identificationi of allergens in extract of horst hair and dandruff by means of crossed radioimmunoelectrophoresis. Lowenstein H, Markussen B and Weeke B. Int Arch Allerg Appl. Immunol. 1976, 51:38.

The role and remediation of animal allergens in allergic diseases. Chapman MD and Wood, RA, M.D. J Allerg Clin Immunol. 2001,107(3):S414.

Health effect of indoor odorants. Cone, JE and Shusterman D. Env. Health Perspec. 1991, 95:53.

Ventilation. Turner W, et al. Occup Med. 1995, 10(1):41.

Air Pollution: The changing perception of health risks, editorial. Indoor+Built Environment. 1997, 6:1.

Sources and Factors Affecting Indoor Emissions from Engineered Wood Products: Saummary and Evaluation of Current Literature. Turner S., et al. June 1996. EPA/600/SR-96/067.

Hair lead levels related to children's classroom Attention-deficit behavior. Tuthill, R. Arch Env Health 1996, 51(3):214.

The effectiveness of a home cleaning intervention strategy in reducing potential dust and lead exposures. Lioy, P, et al. J Expos Anal Env Epid. 1998, 8(1):17.

Contact reactions to fragrances. Katsarou A, et al. Ann Allerg Immunol. 1999, 82:449.

Exposure of children to pollutants in house dust and indoor air. Roberts J and Dickey P. Rev Env Contam Tox. 1995, 143:59.

Enhancing contaminant control to mitigate aeroallergens. Frey A. Ann Allergy Asthma Immunol. 1996, 77:460.

Risky Art Supplies: 13 safe substitutes. Adapted from "Art Materials: Recommendatoins for Children Under 12." Center for Safety in the Arts, Babin, A, Peltz PA, and Rossol M.

The Alert: Allergy to Latex Education and Resource Team, Inc. P.O. Box 23722, Milwaukee, WI 53223-0722.

Formaldehyde: An analysis of its respiratory, cutaneous, and immunologic effects. Bardana Jr. E and Montanaro A. Ann. Allerg. 1991, 66:441.

Is formaldehyde an important cause of allergic respiratory disease? Editorial. Clin Exp Allerg. 1996, 26:247.

Formaldehyde exposure enhances inhalative allergic sensitization in the guinea pig. Hasenauer RF, et al. Allerg. 1996, 51:94.

Formaldehyde. EPA (50-00-0)

The sick bulding syndrome. I. Definition and epidemiological considerations. Chang D, M.D., et al. 1993, 30(4):285.

The sick building syndrome. II. Assessment and regulation of indoor air quality. Chang D, M.D., et al. J Asthma. 1994, 30(4):297.

Building components contributors of the sick building syndrome. Chang, D, M.D., at al. J Asthma. 1994, 31(2)127.

Occupational asthma caused by aromatic herbs. Lemiere C, et al. Allerg. 1996, 51:647.

The free radical basis of air pollution: Focus on ozone. Kelly F, et al. Resp Med 1995, 89:647.

Effect of low concentrations of ozone on inhaled allergen responses in asthmatic subjects. Molfino N, et al. Lancet 1991, 338:1991.

Ozone, airways and allergic airways disease. Krishna MT, et al. Clin Exp Allerg 1995, 25:1150.

Prolonged acute exposure to 0.16 ppm ozone induces eosinophilic airway inflammation in asthmatic subjects with allergies. Peden DB, M.D., et al. J Allerg Clin Immunol 1997, 100(6):802.

Essential Ozone. Alpine Industries. 9199 Central Avenue NE., Minn., Minn.

Pesticides: A review article. Al-Saleh I. J Env Path Tox Onc. 1994, 13(3):151.

Non-Occupational exposures to pesticides for residents of two U.S. Cities. Whitmore RW, et al. Arch Env Contam. Tox. 1994, 26:47.

List of Pesticide Product Inert Ingredients. EPA1995.

Inert Ingredients in Pesticide Products; Policy Statement. EPA / 0PP-36140;FRL-3190-1, 1987.

Inert Ingredients in Pesticide Products; Policy Statement; EPA / 0PP-36140;FRL-3190-1, 1987.

Inert Ingredients in Pesticide Products; Policy Statement; Revision and Modification of Lists. EPA / 0PP-36140;FRL-3190-1, 1989.

Handwipe sampling and analysis procedure for the measurement of dermal contact with pesticides. Geno PW, et al. Arch Environ Contam Tox. 1996, 30:132.

Everyday Exposure to Toxic Pollutants. Ort, WR and Roberts, JW. Scientific Amer. 1998 (Feb):86.

Human exposure to environmental pollutants: A decade of experience. Wallace, LA. U.S. EPA publ. in: Clin Exp Allerg. 1995, 25:4.

Conceptual framework for designing a national survey of human exposure. Lioy PJ and Pellizzari E. J Exp Analysis Env Epid. 1995, 5(3):425.

Scented Candles investigated as cause of indoor soot, staining. Indoor Air Quality Publications. www.iaqpubs.com.

Evaluation of methods for monitoring the potential exposure of small children to pesticides in the residential enviornment. Lewis RG, Fortmann RC and Camann DE. Arch Environ Contam Tox. 1994, 26:37.

Measurement of environmental lead. Flegal AR and Smith DR, In: Rev Env Contam Tox. Springer. 1995. pgs 6-11.

Determination of trihalomethanes produced through the chlorination of water as a function of its humic acid content. Alawi MA, Khalill F and Sahali I. Arch Env Contam Tox. 1994, 26:381.

The effectiveness of a home cleaning intervention strategy in reducing potential dust and lead exposures. Lioy PJ. J Exposure Anal Env Epidem. 1998, 8(1):17.

Clean up of lead in household carpet and floor dust. Ewers L, et al. Amer Ind Hyg Assoc J, 1994, 55:650.

Measurement of deep dust and lead in old carpets. Roberts JW, Glass GL and Spittler TM. In: Measurement of toxic and related pollutants. 1995. Proc Air and Waste Management Assn, San Antonio, TX.

Airborne house dust particles and diesel exhaust particles as allergen carriers. Ormstad H, Johansen BV and Gaarder PI. Clin Exp Allerg. 1998, 28:702.

Effect of two different types of vacuum cleaners on airborne Fel d 1 levels. Ronborg SM, M.D., et al. Ann Allerg Asthma Immun. 1999, 82:307.

Contact reactions to fragrances. Katsarou A, M.D., et al. Ann Allerg Asthma Immun. 1999, 82:449.

Exposure of children to pollutants in house dust and indoor air. Roberts JW and Dickey P. Rev Environ Contam Tox. 1995, 143:59.

Human exposure to pollutants in the floor dust of homes and offices. Roberts JW, et al. J Expos Anal Environ Epid Suppl. 1992, 2:127.

Pesticides: A review article. AL-Saleh, IA. J Environ Path Tox Onc. 1994, 13(3):151.

Guideline to NADCA Standard 1992-01, National Air Duct Cleaners Association, 1518 K Street, N.W., Suite 503, Wash. D.C. 20005.

Comparison of risks from outdoor and indoor exposure to toxic chemicals. Wallace LA. Environ Health Persp. 1991, 95:7.

Chemical contaminanats in house dust: Occurrences and sources. Roberts JW., et al. In: Indoor Air '93: Proceedings of the 6th International Conference on Air Quality and Climate Vol. 2:27. Helsinki University of technology. Espoo, Finland.

Guidelines for control of indoor allergen exposure. Bush RK, M.D. and Eggleston, PA, M.D. J Allerg Clin Immunol. 2001, 107(3):S403.

Environmental allergen avoidance: An overview. Eggleston PA, M.D. and Bush RK, M.D., J Allerg Clin Immunol, 2001,107(3):S403.

On Children and Television. Gotz, IL. Elementary School Journal, April 1975, pp. 415-418.

The scary world of TV's heavy viewer. Gerbner, G and Gross L. Psychology Today, April 1976.

GLOSSARY

AC:
air conditioner (air conditioning)

Allergen:
a particle that can cause an allergic reaction

Allergic:
a person who has an allergy to a particle such as pollen or dust

Allergy:
a disease of the immune system that is characterized by sneezing, itching of eyes, mucous drainage in the throat and sinus headaches; commonly caused by particles or foods

Asthma:
a disease that is characterrized by swelling and constriction of the airways, making breathing difficult

Asthmagenic:
capable of causing symptoms of asthma

Asthmatic:
a person who has asthma

ASHRAE:
American Society of Heating, Refrigeration and Air Conditioning Engineers

CS:
chemical sensitivity

EC:
evaporative cooler

FTC:
Federal Trade Commission

Fumes:
vapors and particles emitted from various substances such as auto exhaust and fires

Fungi:
a large group of microorganisms that include molds and mushrooms

Half-life:

The amount of time it takes one half of a given amount of substance to decay

HEPA:

High efficiency particulate air—a filter that can remove 99.97 percent of particles at a level of 0.3 microns (smaller than most bacteria)

HVAC:

Heating, ventilation and air conditioning

MCS:

Multiple chemical sensitivity

NAS:

National Academy of Sciences

NCEHS:

National Center for Environmental Health Strategies

Molds:

a large group of microorganisms that are present in soil and can contaminate water damaged buildings; are often allergenic

MSDS:

Material Safety Data Sheet

ppm:

parts per million (one part in a million of air or water)

ppb:

parts per billion (one part in a billion of air or water)

Vapors:

gases emitted from various products

VOC:

volatile organic compounds; liquids or solids that give off odorous gases such as glues, solvents, paints, nail polish remover, formaldehyde, ammonia, petroleum distillates

Index

Symbols

A

B

C

D

H

I

J

K

L

M

N

O

P

Q

R

S

T

U

V

W

X

Z

ABOUT THE AUTHOR

Mark Sneller was born in Venice, California in 1942. From 1965-1967 he served in the Peace Corps in India.

Mark Sneller received his Bachelor's Degree in Education from California State University at Los Angeles. He received his Master's Degree from California State University at Long Beach, and Doctorate from the University of Oklahoma, both in Microbiology / Biochemistry with a specialty in Medical Mycology.

He served two post-doctoral appointments at the University of Oklahoma Medical Center in cancer research as well as antibiotic research.

He has taught graduate studies in Medical Mycology at San Jose State University and began an indoor air quality company in Tucson, Arizona in 1979.

Since that date, Dr. Sneller has received two awards from the Arizona Lung Association for work in the field of respiratory health, has been featured in Newsweek Magazine, the New York Times, ABC, NBC and CBS national network news. Dr. Sneller, has also been under contract with the Department of Justice and Department of Defense and served on the State of Arizona Air Pollution Control Hearing Board under appointment from the governor. He helped institute and oversaw the nation's first pollen control law, has hosted his own radio talk show and has been a weekly newspaper columnist for the past ten years. He is currently a bioterrorism consultant to the City of New York, Department of Health and Mental Hygiene.

Dr. Sneller is an approved pollen and mold identification expert by the American Academy of Allergy, Asthma and Immunology and is the author of some fifteen scientific papers in the fields of mycology, palynology, organic chemistry, fungal toxins and combination drug therapy.

Dr. Sneller is a Sensei with the Japan Karate Association and is a member of the Society of American Magicians and the Society of Southwestern Authors.